W9-CDO-543

THE RUBY & SAPPHIRE BUYING GUIDE

How to Spot Value & Avoid Ripoffs

Text & Photographs

by

RENEE NEWMAN

International Jewelry Publications

Los Angeles

This publication is designed to provide information in regard to the subject matter covered. It is sold with the understanding that the publisher and author are not engaged in rendering legal, financial, or other professional services. If legal or other expert assistance is required, the services of a competent professional should be sought. International Jewelry Publications and the author shall have neither liability nor responsibility to any person or entity with respect to any loss or damage caused or alleged to be caused directly or indirectly by the information contained in this book. All inquiries should be directed to:

International Jewelry Publications
P.O. Box 13384
Los Angeles, CA 90013-0384 USA

(Inquiries should be accompanied by a self-addressed, stamped envelope).

Printed in the United States of America

10 9 8 7 6 5 4 3 2 1

Library of Congress Cataloging in Publication Data

Newman, Renée.
The ruby & sapphire buying guide : how to spot value & avoid ripoffs / text & photographs by Renée Newman.
p. cm.
Includes bibliographical references (p.) and index.
ISBN 0-929975-15-4 : $19.95
1. Rubies--Purchasing. 2. Sapphires--Purchasing. I. Title. II. Title: Ruby and sapphire buying guide.
TS755.R82N48 1990
736'.25--dc20 90-83178
CIP

Cover photograph by
Bobbi Lane and Renée Newman
The 5.22 carat heart-shaped sapphire is courtesy of the Josam Diamond Trading Corporation.
The oval stone is a 1.73 carat lab-grown ruby. Its retail cost--$4.

Contents

Preface

That jeweler you recommended sold me quartz, not sapphire.

When I was a tour director, one of my passengers phoned me with this complaint. She had purchased an expensive yellow sapphire necklace in Bangkok from a jeweler I knew.

After she got back to the United States from the Orient tour I'd conducted, she took the necklace to an appraiser. He identified all of the yellow stones in it as citrine quartz, so the appraised value of the necklace was much less than she had paid.

This surprised me because the jeweler that had sold her the necklace was a highly qualified gemologist with an excellent reputation. Therefore, I suggested we first get an explanation from the jeweler in Bangkok before returning the necklace.

The jeweler wrote back that the yellow stones were sapphires, and he outlined several tests for the appraiser to do. He went on to assure her that she needn't worry about getting a full refund.

The tests showed that the yellow stones were, indeed, sapphires; plus, the resultant appraised value of the necklace was much greater than the passenger had paid. The passenger was happy because she had a one-of-a-kind jewelry piece, and she had wanted to keep it. In fact, the appraiser was probably so surprised to see such an unusual sapphire necklace that he assumed the stones to be citrine.

At the time this incident occurred, I had started taking courses at the GIA (Gemological Institute of America), which eventually led to a graduate gemologist diploma and my entry into the jewelry trade. Thanks to what I had learned at the GIA, I had enough knowledge at the time to understand the qualifications of the Bangkok jeweler and to realize how well-informed he was about sapphires. Consequently, I didn't jump to any false conclusions.

The statement "A little knowledge is a dangerous thing" is often used as an excuse to learn nothing at all about gemstones. One thing is certain--when it comes to buying gems, knowing nothing about them is far more dangerous than knowing something. How can consumers, for example, intelligently evaluate the qualifications of a jeweler or appraiser if they know nothing about gems and trade credentials?

Rubies and sapphires can be a major investment, and you can't afford to make the wrong choice or buy from an unreliable person. With the proper knowledge and professional help, you *can* make the right choices. Until now, there have been no books that offer laypeople detailed, step-by-step instructions on how to judge the quality of rubies and sapphires. I wrote this book to fill that void. Use it along with the advice of your jeweler to select the rubies and sapphires that will suit both your pocketbook and your needs.

Acknowledgements

I would like to express my appreciation to the following people for their contribution to *The Ruby & Sapphire Buying Guide*:

Ernie and Regina Goldberger of the Josam Diamond Trading Corporation. This book could never have been written without the experience and knowledge I gained from working with them. Most of the rubies and sapphires pictured in this book are or were part of their collection.

The American Gemological Laboratories and the Gemological Institute of America. They have contributed diagrams, photos, and information.

Jeffrey Badler, Jeffrey Baer, C. R. Beesley, Shirley Bradshaw, Charles Carmona, Dona Dirlam, Louise Harris, Susan B. Johnson, Ronny Levy, Peter Malnekoff, Linda Newton, Howard Rubin, Sindi Schloss, and Tom Tashey. They have made valuable suggestions, corrections, and comments regarding the portions of the book they examined. They are not responsible for any possible errors, nor do they necessarily endorse the material contained in this book.

Les Gemmes d'Orient, Inc., Overland Gems, Inc., and Tory Jewelry Company, all in Los Angeles. Their stones have been used for some of the photographs.

Grieger's Inc., the Gubelin Gemological Laboratory, and the Smithsonian Institution. Material from them has been reproduced in this book.

John Budd, Bobbi Lane, and George Seifert. They have provided technical assistance.

Patricia S. Esparza. Besides spending hours carefully editing *The Ruby & Sapphire Buying Guide*, she has tried out the identification tests. Thanks to her, this book is much easier for consumers to read and understand.

My sincere thanks to all of these contributors for their kindness and help.

1

Why Read a Whole Book Just to Buy a Ruby or Sapphire?

A Thai massage and a ruby ring. Don was determined to get both of these before leaving Bangkok. Since Thailand was the ruby capital of the world, Don figured that he could buy a fine intense-red ruby there for a couple hundred dollars. He was stunned to find jewelers in Bangkok asking several thousand dollars for such stones.

One day as Don was leaving his hotel, a well-dressed gentleman approached him. As he showed Don his card and badge he said, "Sir, I represent the Ministry of Tourism. We've been investigating a lot of the jewelry stores and lapidaries in this area that cater to foreigners, and we've discovered they are overcharging them. We think you should know that if you shop where the locals shop, you'll pay a lot less."

Don then asked him if he knew where a person could get a good deal on a ruby ring. The man said he did and proceeded to take Don to a small shop on one of the back streets.

There was an amazing difference in the prices. In fact, Don was able to find an attractive 1 3/4 carat ruby ring for just $450. He couldn't pass it up, especially since the ruby didn't seem to have any flaws, and it even came with a certificate of guarantee. Don gave the man from the Ministry of Tourism a twenty-dollar tip for helping him save so much money.

When Don got back home, he couldn't wait to take his ruby ring to his appraiser and find out how much it was worth. To his dismay, she told him that the ring was gold-plated and that the ruby was an inexpensive type of man-made ruby. The total value of the ring was about $10. There was no way for Don to get his money back. The certificate of guarantee was just as phony as the badge and identity card of the well-dressed man.

Bill and Linda were on vacation in Thailand. Linda had always wanted a sapphire pendant; since their anniversary was coming up in a couple weeks, Bill figured this was the perfect time to get her one. They went to a jeweler Bill had met on a previous trip and Linda picked out a pear-shaped sapphire surrounded by diamonds.

After leaving Thailand, they had a one-day layover in another country. While they were browsing in a jewelry store there, the salesman said to Linda "That's a beautiful pendant you're wearing. I bet you bought it in Thailand."

She nodded yes, and the salesman asked to have a closer look at it. After examining it with his jeweler's magnifying glass, he shook his head and said "Just as I thought. Those jewelers in Thailand are nothing but a bunch of thieves. This sapphire has a crack in the corner."

Bill and Linda worried about the sapphire for the rest of the trip home. Even though they knew they could get their money back, they didn't want to have to go through the hassle of returning it. Besides, the pendant already had sentimental value to them.

When they got back home, they took it to their jeweler. He told them that the color and overall quality of the sapphire were exceptional. They were surprised and asked him about the crack. He said it was a normal flaw and it was so minute they had nothing to worry about. Then he went on to explain how flaws can help prove a sapphire is natural and how they can sometimes even increase a sapphire's value by revealing the country it is from. Bill and Linda were relieved, but they had gone through a lot of needless worry.

Jesse was looking for a sapphire engagement ring for his girlfriend Patty. Blue was Patty's favorite color, and Jesse had discovered that even though sapphires weren't cheap, they cost a lot less than diamonds. After looking at several rings, he picked out one with a one-carat oval sapphire.

Patty was pleasantly surprised when Jesse slipped the ring on her finger. She had always wanted a sapphire. As she was looking at it, however, she noticed she could see the pores of her finger right through the stone. Jesse was disappointed that the color looked much lighter than it had in the store. Even though Patty pretended she was happy with the ring, she wished Jesse would have chosen something else.

Divonna was about to leave India when a peddler asked her if she'd like to see a star ruby. Out of curiosity, she decided to have a look. It was grayish purple, weighed four carats, and had a fairly distinct star. Divonna was amazed to find out it cost only $150 because she had heard star rubies were worth a fortune. There was no way she was going to pass up a deal like this. With all the money she'd make selling this ruby, she figured she could pay for a trip to Colombia where she'd buy some emeralds and really strike it rich.

When Divonna got back home and showed the stone to her jeweler, he confirmed that it was a genuine Indian star ruby. However, he said that due to the color and the fact that no light shone through the stone, it wasn't worth much. He proceeded to tell her that he'd seen some star rubies of the same size and quality at a recent gem show. Their price--$4 a carat.

All of the negative experiences in the above stories could have been avoided. Suppose Divonna had had a book that explained how star rubies and sapphires were valued. Wouldn't it have helped her realize that the star ruby she was offered was over-priced?

Suppose Jesse had had a book that showed how to judge sapphire quality. Couldn't it have helped him select a stone that would be more appealing to Patty?

Suppose Bill and Linda had had a book that described the normal types of flaws found in sapphires. Wouldn't this have helped them avoid needless worry when they discovered such flaws in their own sapphire? If the book had also described unacceptable flaws, couldn't it have helped them avoid regrets in future sapphire purchases?

Suppose Don had had a book that warned him about man-made rubies being sold as natural stones. And suppose this book had explained that even in Thailand, gem dealers don't sell extra-fine rubies at prices way below market value. Wouldn't it have helped him be suspicious about such a low-priced ruby?

If you glance at the table of contents of *The Ruby & Sapphire Buying Guide*, you'll notice a wide range of subjects relevant to buying rubies and sapphires. There is no way a brochure could cover these subjects adequately. Likewise, it would be impossible for jewelers to discuss thoroughly the grading, identification, pricing, and enhancement of colored stones during a brief visit to their store. It would be better to first learn some fundamental information by reading this book. Jewelers can show you how to apply your new-found knowledge when selecting a stone, and they can help you find what you want.

What This Book Is Not

♦ It's not a guide to making a fortune on rubies and sapphires. Nobody can guarantee that these stones will increase in value and that they can be resold for more than their retail cost. However, understanding the value concepts discussed in this book can increase your chances of finding good buys on rubies and sapphires.

♦ It's not a ten-minute guide to appraising rubies and sapphires. There's a lot to learn before being able to accurately compare these stones for value. That's why a book is needed on the subject. *The Ruby & Sapphire Buying Guide* is just an introduction, but it does have enough information to give laypeople a good background for understanding price differences.

♦ It's not a scientific treatise on the chemistry, crystallography, and geological distribution of rubies and sapphires. The material in this book, however, is based on technical research; the appendix lists the physical and optical properties of these stones to help you identify them. Technical terms needed for buying or grading colored stones are explained in everyday language.

- It's not a discussion about the mining and prospecting of rubies and sapphires. There is a whole chapter, however, that outlines where they are found because this can have an affect both on their price and their appearance.

- It's not a substitute for examining actual stones. Photographs do not accurately reproduce color, nor do they show the three-dimensional nature of gemstones very well. Concepts such as brilliancy and transparency are best understood when looking at real stones.

What This Book Is

- A guide to evaluating the quality of rubies and sapphires.

- An aid to avoiding fraud with tips on detecting imitations and synthetic stones.

- A handy reference on rubies and sapphires for laypeople and professionals.

- A guide to choosing appraisers and gem labs.

- A collection of practical tips for travelers buying gems.

- A challenge to view rubies and sapphires through the eyes of gemologists and gem dealers.

How to Use This Book

The Ruby & Sapphire Buying Guide is not meant to be read like a murder mystery or a romance novel. Some laypeople may find this book overwhelming at first. It might be advisable for them to start by looking at the pictures and by reading Chapter 2 (Curious Facts about Rubies & Sapphires), Chapter 19 (Finding a Good Buy), and the Table of Contents. Then they should learn the basic terminology in Chapter 3 and continue slowly, perhaps a chapter at a time.

Skip over any sections that don't interest you or that are too difficult. This book has far more information than the average person will care to learn. That's because it's also designed to be a reference. When questions arise about rubies and sapphires, you can avoid lengthy research by having the answers right at your fingertips.

To get the most out of *The Ruby & Sapphire Buying Guide*, you should try to actively use what you learn. Buy or borrow a loupe (jeweler's magnifying glass) and start examining any jewelry you might have at home. Take the quizzes that you'll find at the end of many of the chapters. Look around in jewelry stores and ask the professionals there to show you different qualities and varieties of rubies and sapphires. If you have appraisals or grading reports, study them carefully. If there is something you don't understand, ask for an explanation. When you examine jewelry, keep in mind that diamonds and other gemstones are not valued in the same way as the ruby and the sapphire. However, knowing how to grade these two stones will make it easier for you to grade other gemstones.

Shopping for rubies and sapphires should not be a chore. It should be fun. There is no fun, though, in worrying about being deceived or in buying a stone that turns out to be a poor choice. Use this book to gain the knowledge, confidence, and independence needed to select the stones that are best for you. Buying gemstones represents a significant investment of time and money. Let *The Ruby & Sapphire Buying Guide* help make this investment a pleasurable and rewarding experience.

2

Curious Facts About Rubies & Sapphires

An Informal Message from the Ruby and Sapphire Family:

We rubies and sapphires may look very different, but we're from the same family. We're all just a combination of aluminum and oxygen (Al_2O_3).

Even though our family name is Corundum, we prefer to be called by our first names. Our names are easy to learn. If we're red, we're called rubies; if we're any other color, we're called sapphires.

Believe it or not, we come in just about every color imaginable--green, blue, black, orange, pink, brown, gray, yellow, purple, and we can even be colorless. Occasionally, we're bi-colored or multi-colored. Some of us change color when we go outdoors in the sun. Others look like colored stars. To avoid confusion, we often include our color with our name, as in "yellow sapphire."

Our family has a lot more to be proud of than just our diversity. For centuries we have been considered regal gems. In ancient India, the ruby was called "king of gems." In England, the ruby was used for coronation rings. Sapphires were often worn by kings and queens around the neck for good luck. Considering the high regard British royalty has had for our family, it's not surprising that Princesses Anne and Diana each received a sapphire engagement ring and Fergie the Duchess of York received one with a ruby. Our presence in engagement rings hasn't been limited to royalty. Liz Taylor and Luci Baines Johnson Nugent (daughter of US president Lyndon Johnson) are two other well-known women that have received sapphire engagement rings.

We are not only royal gems, we are sacred as well. In the Catholic church, sapphires have been used in the rings of bishops and cardinals. Our blue color symbolizes heaven; and supposedly, people who wear us become more virtuous, devout, and wise.

Buddhists believed sapphires signified friendship and steadfastness. Ancient Hindus thought if they offered a ruby to the god Krishna, they would be reborn as an emperor. According to Hindu writings, the ruby represented the sun; and the sapphire, the planet Saturn.

Perhaps you're wondering why we have been so revered throughout history. It's partly because of our rich looking colors. Thanks to them, we can look like jewels even when we're cut into simple rounded stones with no geometric facets. A diamond cut this plain would look rather drab.

Many people don't realize that we are each a blend of two colors. For example, in one direction a ruby looks purplish and in another, orangy. But when you view it as a whole, you see a sumptuous red.

To top it all off, the ruby is blessed with a red aura. It's usually in the sun that you see this distinctive glow. But the ancient Burmese said it could even be seen in the dark. According to one legend, a king in Burma had rubies that glowed so brightly, they lit up the city at night.

Rubies have used their color to full advantage. They grab your attention with it, and then cleverly manipulate you into thinking they are bigger and closer than they actually are. This helps rubies compensate for the fact that their average size is generally less than other gemstones. Their small size is not necessarily a detriment. They use it to prove they are very rare. Consequently, rubies can bring unusually high prices. Believe it or not, in 1988, a 15.97 carat ruby sold for $3,630,000 (per carat cost, $227,300). The only diamonds that have brought such high per-carat prices are those which have colors like ours. Our friends the diamonds brag about their rarity. Our family, though, is far more rare. That's another reason we're so revered.

Diamonds do have something over us. They are several times harder. But no other gemstones are harder than we are. Actually, our family is glad we're number 2 in terms of hardness. This means that diamonds get stuck with the majority of the cutting, grinding, and polishing work. Some of it, though, is still reserved for us. Plus our man-made brothers and sisters serve as styluses in record players, tips in ballpoint pens, and jewel bearings in watches, meters, and aircraft instruments. Since 1960, we've been used as a core in lasers.

We've even been used as doorstops. An Australian gem buyer paid around $24 for a rough sapphire about the size of a chicken egg (1156 carats). It had originally been found by a little boy and used as a doorstop. Later, the gem buyer sold the stone to an American who cut it into a 733-carat star sapphire called the "Black Star of Queensland." According to one report, it's value now is estimated at over $267,000.

When we're as big as the "Black Star of Queensland," we may be impractical to wear as jewelry, so sometimes we're carved into attractive figurines. A select few of us are sculptures of famous people like Confucius, George Washington, and Dr. Martin Luther King. Confucius was carved from a multi-colored sapphire in such a way that his head is white, his trunk and arms blue, and his legs yellow. During a period of one and a half years, Washington was carved from a blue sapphire into a 1056-carat sculpture. Dr. King was carved from a rough sapphire weighing 4180 carats, and the final 3294-carat sculpture of him was unveiled in 1984.

It's an honor for us to represent such great men. It's an honor for us to be so revered by churches and royalty. It's also an honor when you wear us and appreciate us. Our family has a lot to offer you--beauty, strength, variety, mystery, romance. So please, take us home with you. Let us add some more color to your life.

3

Shape & Cutting Style

When buying a diamond, choosing the shape is often the first thing to consider. When buying a ruby or a sapphire, it's often one of the last considerations. For example, when a person finds a brilliant, one-carat ruby that's just the color he or she is looking for, there's a tendency to accept whatever shape it has because it's so hard to find a large, well-cut ruby with the desired color.

In sizes of a carat or more, rubies and sapphires usually have either an oval or **antique cushion shape** (square or rectangular with curved corners and sides. It's often just called a **cushion** by gem dealers)(fig. 3.1). Cushions and ovals are the two shapes that usually allow cutters to save the most weight of the original rough. However, when rough crystals have other shapes like triangles, kites, shields, pears, etc., they can be cut in these forms as well (figs. 3.11 to 3.15). You may notice that the shape of rubies and sapphires is often not as symmetrical as that of less-expensive colored stones. That's because cutters know a lot of money can be lost when rubies and sapphires are cut down to symmetrical shapes. On the other hand, very little money is lost when stone such as a blue topaz is cut symmetrical. In fact, this improves the look of the topaz, and it can be calibrated to specific sizes and sold in large quantities for mass-produced jewelry.

In smaller sizes, say less than 1/2 carat, rubies and sapphires are commonly cut into the traditional round, oval, pear, marquise, square, and rectangular shapes. Consequently, buyers can usually find the shape they want when looking for small stones.

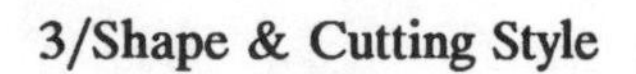

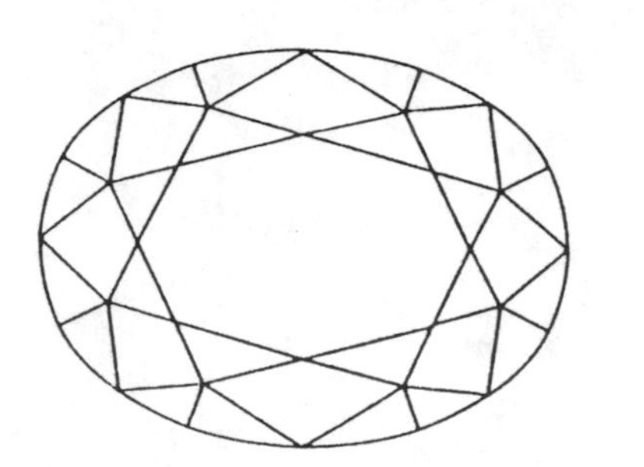
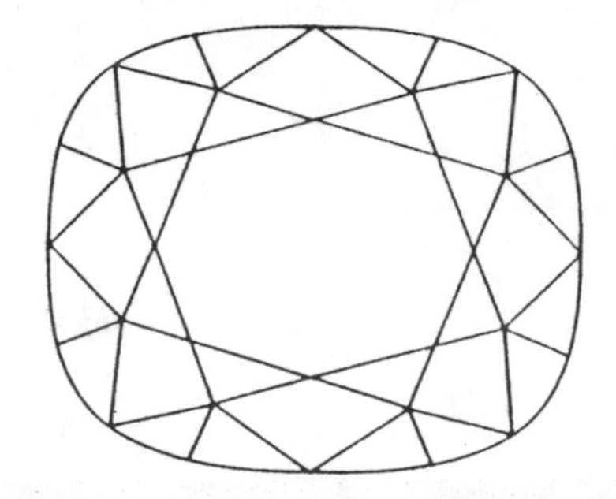
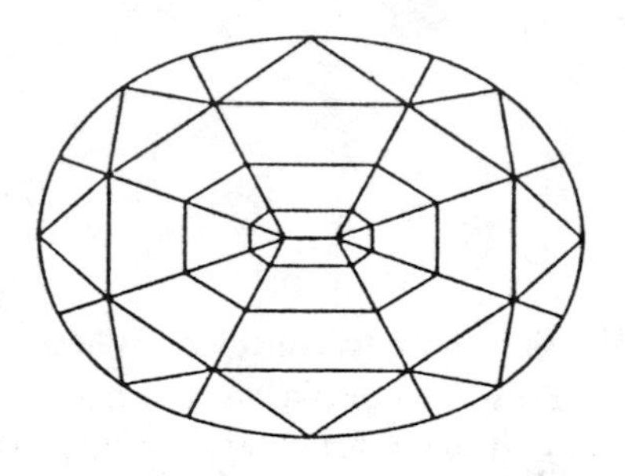
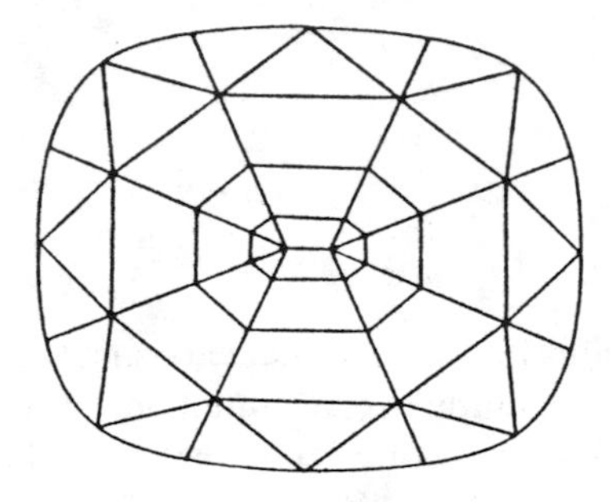
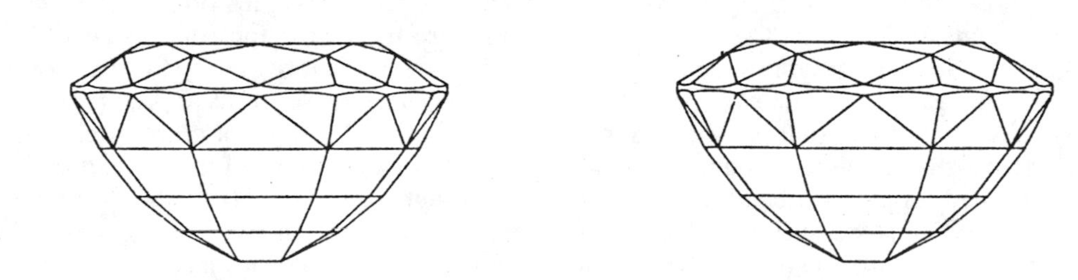

Fig. 3.1 Oval and cushion mixed cuts. Copyright 1982 by American Gemological Laboratories.

Gemstone Terms Defined

Before you can thoroughly understand a discussion of shapes and cutting styles, some terminology must be explained. A few basic terms are described below and illustrated in figure 3.2.

Facets — The flat, polished surfaces or planes on a stone.

Table — The large, flat top facet. It normally has an octagonal shape on a round stone,

Girdle — The narrow rim around the stone. The girdle plane is parallel to the table and is the largest diameter of any part of the stone.

Crown — The upper part of the stone above the girdle.

Pavilion — The lower part of the stone below the girdle.

Culet — The tiny facet on the pointed bottom of the pavilion, parallel to the table. Sometimes the point of a stone is called "the culet" even if no culet facet is present, which is usually the case with rubies and sapphires.

Fancy Shape — Any shape except round. This term is most frequently applied to diamonds.

Cutting Styles

Before the 1300's, gems were usually cut into unfaceted rounded beads or into cabochons (unfaceted dome-shaped stones). Colored stones like rubies and sapphires looked attractive cut this way, but diamonds looked dull. It's thanks to man's interest in bringing out the beauty of diamonds that the art of faceting gemstones was developed. At first, facets were added haphazardly, but by around 1450, diamonds began to be cut with a symmetrical arrangement of facets. Various styles gradually evolved, and by the 1920's, the modern round-brilliant cut was popular.

As cutters discovered how faceting could bring out the brilliance and sparkle of diamonds, they started to apply the same techniques to colored stones. Today, rubies and sapphires are cut into the following styles:

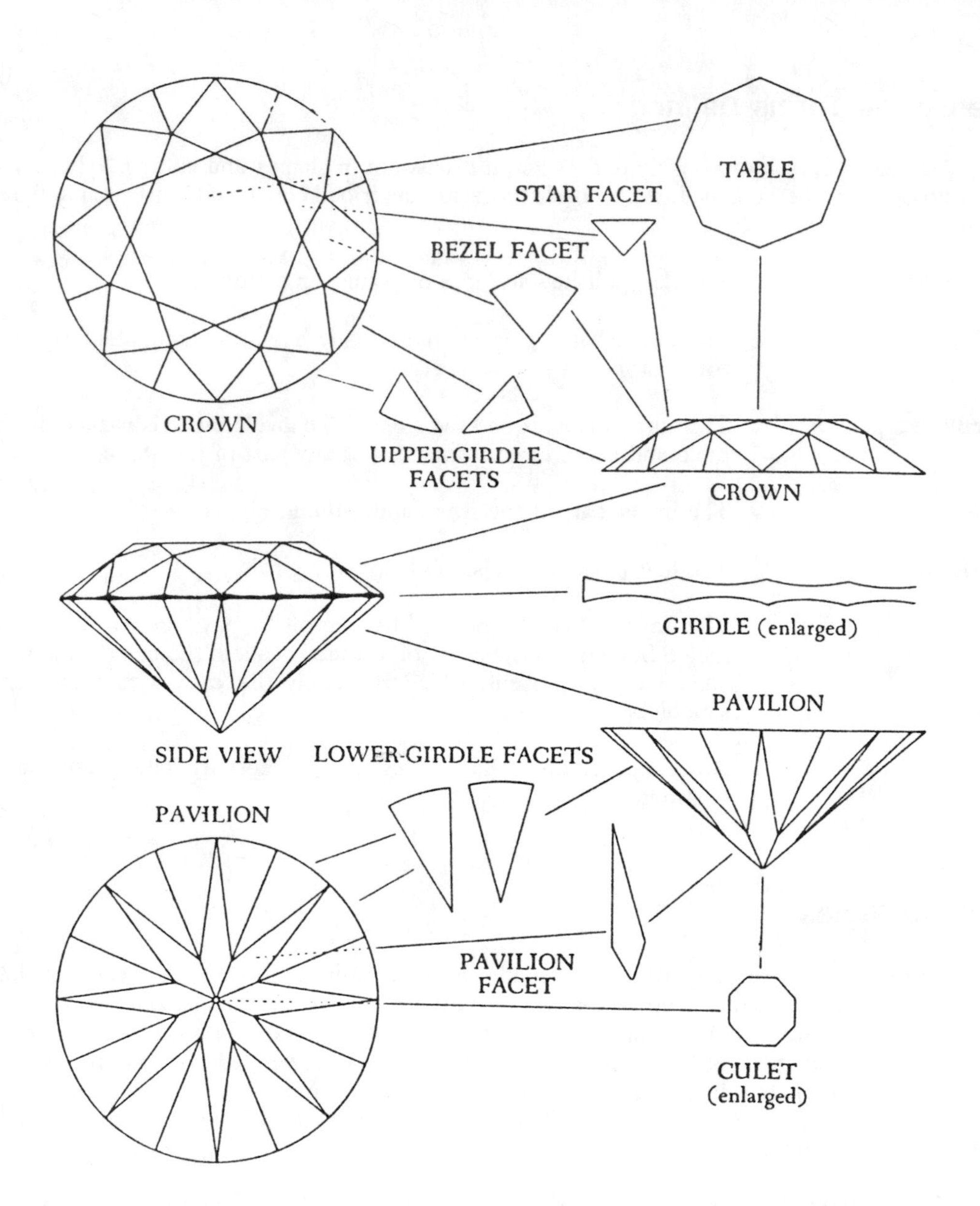

Fig. 3.2 Facet arrangement of a standard round brilliant cut. Diagram courtesy of the GIA.

Cabochon Cut

Has a dome-shaped top and either a flat or rounded bottom (fig. 3.17). This is the simplest cut for a stone and is often seen in antique jewelry. Today this cut tends to be used for opaque, translucent, and star & cat's-eye stones. Transparent rubies and sapphires with lots of flaws are also frequently cut into cabochons.

Step Cut

Has rows of facets that are cut parallel to the edges and resemble the steps of a staircase. Corundum (ruby & sapphire) is commonly fashioned into small step-cut **squares** or **baguettes** (square-cornered, rectangular stones) (fig. 3.3). If step-cuts have corners that look as if they were clipped off, they're called **emerald cuts** since emeralds are typically cut this way. Natural rubies and sapphires weighing a carat or more are seldom cut in an emerald style.

Brilliant Cut

Has triangular-, kite-, or lozenge-shaped facets that radiate outward around the stone. The best-known example is the **full-cut round brilliant**, which has 58 facets (fig. 3.2). Another example, is the **single cut**, which has 17 or 18 facets and is used on small stones that are often of low quality (fig. 3.18). On rubies and sapphires, however, the number of facets can vary, even on round stones. Square rubies and sapphires cut in the brilliant style are called **princess cuts** (fig. 3.19). They may also have a trademarked name such as "Q-Brilliant." The number of facets on these corundum stones varies, but usually it ranges from 35 to 45 facets. Triangular brilliant cuts are called **trilliants**. The princess and trilliant cuts were originally developed for diamonds because their brilliant-style facets create a greater amount of brilliance and sparkle than step facets do. Now the princess and trilliant cuts are becoming popular for small corundum stones. Even though most rubies and sapphires of a carat or more have some brilliant-type facets, these larger stones are not commonly cut in a full-brilliant style. Cuts such as the step and mixed cut intensify their color more effectively.

Mixed cut

Has both step- and brilliant-cut facets. For corundum, this is the most common faceting style. Usually the crown is brilliant cut in order to maximize brilliance and sparkle. The pavilion on the other hand is either entirely step cut or else has a combination of both step-and brilliant-type facets. The step facets allow cutters to save weight and bring out the color of the stone. Occasionally, the mixed cut is referred to as the **Ceylon cut**.

Bead (faceted & unfaceted)

Usually has a ball-shaped form with a hole through the center. Most faceted beads have either brilliant- or step-type facets (figs. 3.9 and 3.10). Corundum beads are generally made from nontransparent or heavily flawed material.

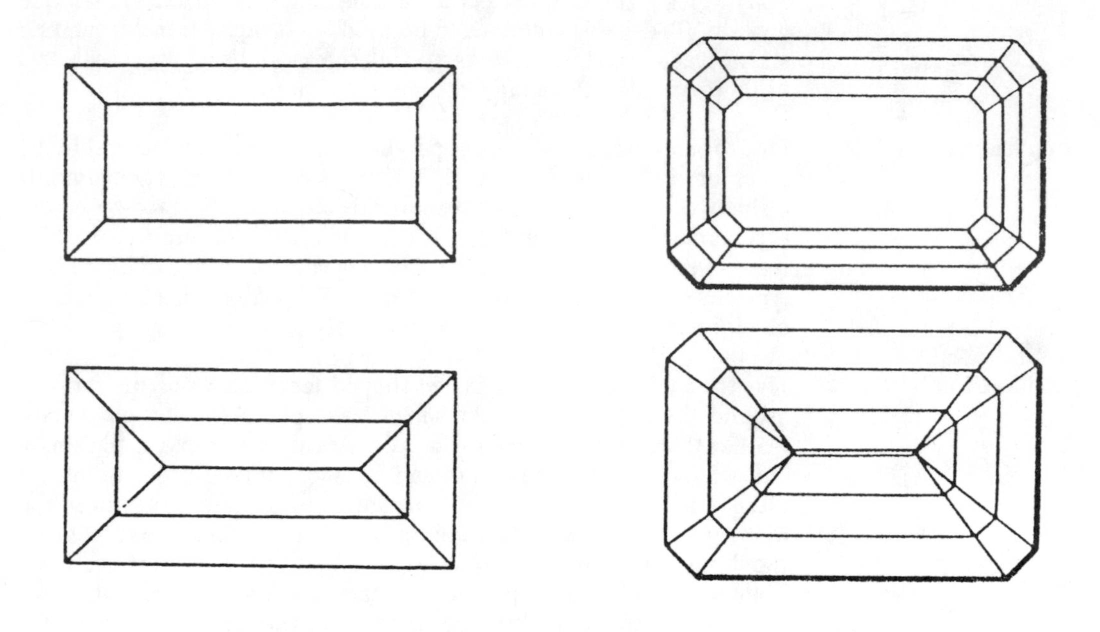

Fig. 3.3 Baguette

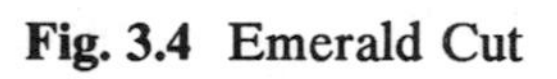

Fig. 3.4 Emerald Cut

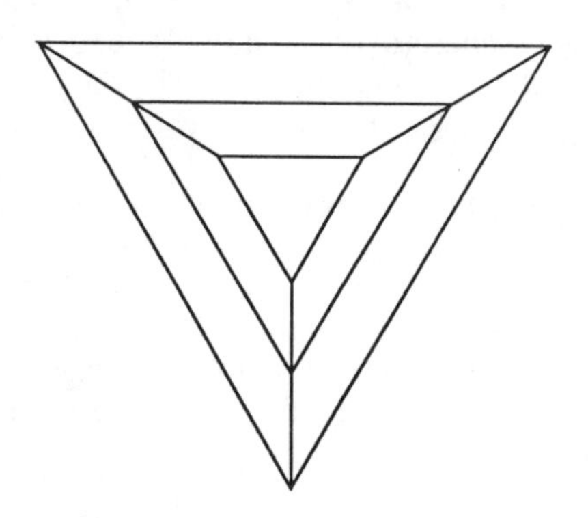

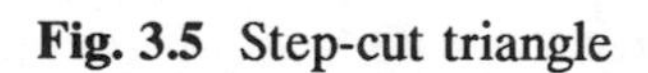

Fig. 3.5 Step-cut triangle

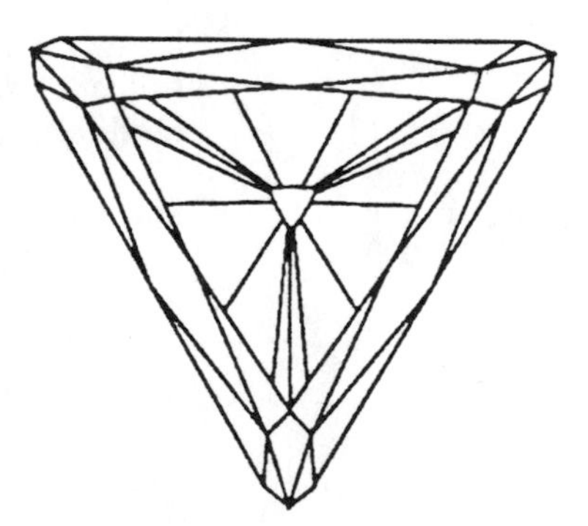

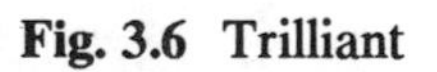

Fig. 3.6 Trilliant

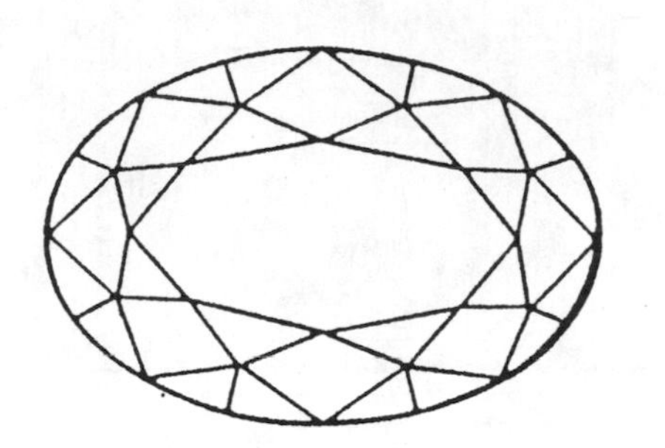

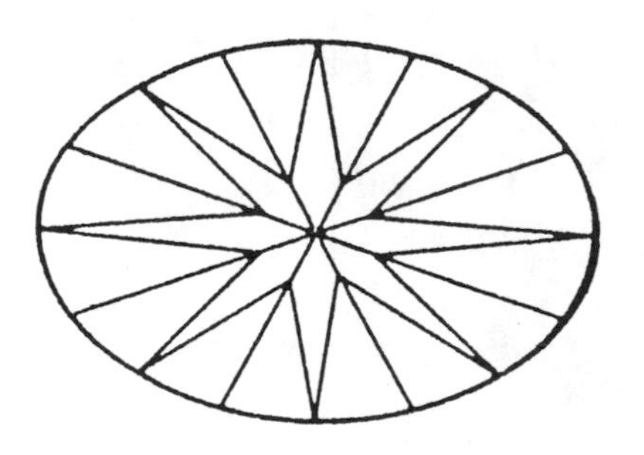

Fig. 3.7 Oval brilliant cut

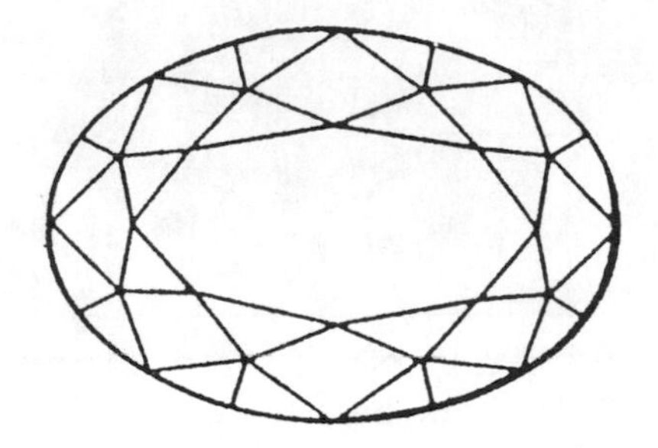

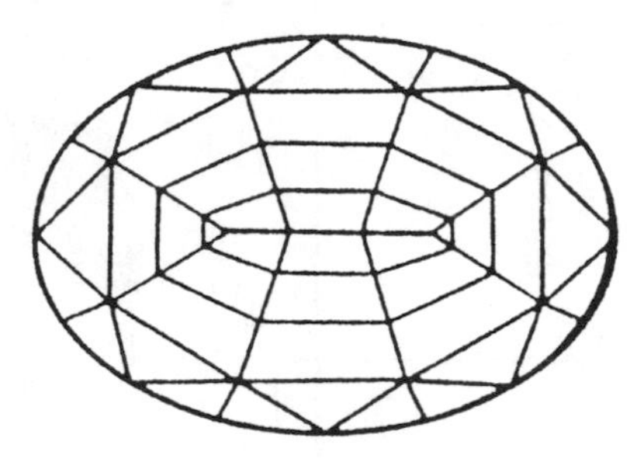

Fig. 3.8 Oval mixed cut

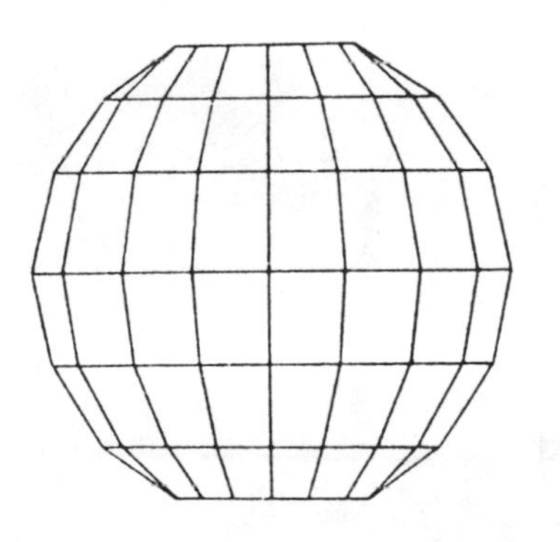

Fig. 3.9 Step-cut bead

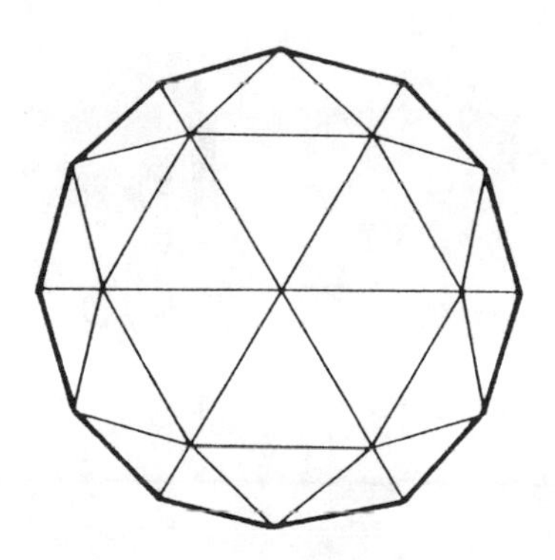

Fig. 3.10 Bead with brilliant-style facets

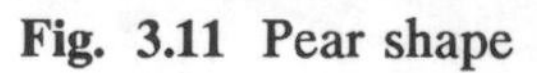

Fig. 3.11 Pear shape

Fig. 3.12 Oval

Fig. 3.13 Marquise

Fig. 3.14 Kite shape

Fig. 3.15 Half-moon shape

Fantasy Cut

Has a free form and is a one-of-a-kind sculpture (fig. 3.16). The design may be geometric or abstract, or it can represent something concrete like a fish or flower. Gem-quality corundum is not often carved into fantasy cuts. Due to the high cost and hardness of corundum, gem materials such as topaz, tourmaline, amethyst, and citrine quartz are preferred for these cuts.

Fig. 3.16 An abstract fantasy cut

How Shape & Cutting Style Affect Price

Color, clarity, & brilliance normally play a greater role in determining the price of a ruby or sapphire than shape and cutting style. Nevertheless, these two factors can affect the value of these stones. The way they affect it is described below.

Since the **cabochon** is the simplest style, it costs less to cut than faceted styles. Another reason cabochons are generally priced less is that they are often made from lower quality material that is unsuitable for faceting. Cabochon rubies and sapphires can also be of high quality, especially those found in antique jewelry.

The shape of **faceted** rubies and sapphires of one carat or more usually has a greater effect on their price than the faceting style. Corundum dealers, however, do not always agree as to which shapes are the most valuable. This is illustrated by comparing shape/price correlations outlined in *The Guide* and the *GAA Precious Gem Market Monitor*, two publications that offer colored-stone pricing information based on dealer surveys.

Fig. 3.17 Cabochon

Fig. 3.18 Low-quality, single-cut ruby

Fig. 3.19 Princess-cut yellow sapphire

Fig. 3.20 Trilliant-cut yellow sapphire

According to the 1990-1991 *Guide*, the shape/price correlations for good to extra-fine rubies and blue sapphires are **generally** as follows:

marquise or emerald cuts	5% to 20% more than cushions & ovals
rounds	0% to 20% more
pear shapes	0% to 15% more

(The better the stone, the higher the percentage of increase)

According to the current *GAA Precious Gem Market Monitor* the shape/price correlations for corundum are **generally** as follows:

<u>Rubies</u>

rounds	10% to 20% more than cushions & ovals
emerald cuts	same or a bit more
marquise & pear shapes	5% to 20% less

<u>Sapphires</u>

rounds & emerald cuts	10% more than cushions & ovals
marquise & pear shapes	same as cushions & ovals

Perhaps you're wondering why the results of these two surveys are so different. One reason is that each publication got its information from a different group of dealers. Another reason is that many dealers don't have a lot of large pear, marquise-, and emerald-cut stones from which to make shape/price comparisons. The majority of their rubies and sapphires of a carat or more are cushions and ovals. Both *The Guide* and the *GAA Gem Market Monitor* acknowledge that sharp variations of pricing may exist from one dealer to another. Colored stone pricing is not as standardized as diamond pricing.

Two generalizations that can be made from these two surveys and from talking to other dealers are:

- Shape can affect the price. Consequently, when you judge prices, you should compare stones with the same shape.

- Round rubies and sapphires of a carat or more generally cost more than cushions and ovals, particularly in the higher qualities. (The shape of small stones usually has little effect on their price.) Dealers explain that large round stones often cost more than other shapes because they have to be cut more precisely to look good. This means that normally more weight is lost from the rough during cutting.

There are cases, however, where round stones may cost a lot less than an oval because there may not be a demand for them. According to one corundum dealer, if he had a large, oval Burmese ruby, he could ask twice as much for it as for a round one because it would be easier to find a buyer for the oval stone.

Beads, even when they are faceted, are generally priced lower than other cuts because they tend to be made of inferior material. It wouldn't make sense to decrease the weight and value of fine quality corundum by drilling holes through it. Consequently, corundum beads are normally made from nontransparent or heavily flawed material that isn't suitable for other cuts.

Fantasy cuts are often priced per piece instead of per carat. Their value varies according to:

- the time spent planning and executing their design.
- the skill and fame of the cutter.
- the quality of the material used. In the case of corundum carvings, cutters tend to use lower-priced nontransparent material. Higher quality material may be used for faceted fantasy cuts.

It would be pointless to contrast the prices of fantasy-styles to the prices of the other cuts. Each fantasy-cut stone should be judged on its own artistic merits. When judging the prices of other cuts, keep in mind that it's best to compare stones of the same cutting style as well as the same shape, size, color, and clarity. This will help you judge value more accurately.

Quiz (Chapters 2 and 3)

True or False?

1. Rubies and sapphires tend to be cut less symmetrical than stones such as amethysts and blue topaz.

2. The girdle of a gem is its smallest width.

3. Cushion-shaped rubies and sapphires have more facets and usually cost more to cut than ovals and rounds.

4. The table is the large flat top surface of a gem.

5. The mixed cut is the most common faceting style for rubies and sapphires.

6. Diamonds are more rare than rubies.

7. There is agreement among dealers as to what shape rubies and sapphires are the most valuable.

8. Red corundum is a technical term for ruby.

9. Sapphires have the same chemical composition and physical properties as rubies.

10. Sapphires are harder than jade and steel.

11. The top part of a gemstonc is called the pavilion.

12. Princess cuts, trilliants, and single cuts are brilliant cuts.

13. Large sapphires are easier to find than large rubies.

14. The cabochon cut is usually reserved for star stones and the finest quality rubies and sapphires.

15. When judging prices, consumers should try to compare stones of the same shape and cutting style.

Answers:

1. True

2. False - The girdle of a gem is the outer edge where the width, length, or diameter is the largest.

3. False - Cushion-shaped rubies and sapphires don't usually cost more to cut and they may have fewer, more, or the same number of facets as ovals and rounds. The number of facets on rubies and sapphires is variable.

4. True

5. True

6. False

7. False

8. True

9. True

10. True - With the exception of the diamond, no gemstone or metal is harder than the ruby or sapphire. Corundum is over 700 times harder than jade and steel.

11. False - The top part above the girdle is the crown. The pavilion is the lower portion.

12. True

13. True

14. False - The finest quality transparent rubies and sapphires are generally faceted. The lower qualities are often cut as cabochons.

15. True

4

Judging Ruby Color

A clear, transparent, and faultless ruby of a uniform deep red colour is at the present time the most valuable precious stone known. (1896, Dr. Max Bauer, German mineralogist, from his book *Precious Stones*.)

Red diamonds have now surpassed rubies in value, but Dr. Bauer's description of the most prized ruby color is still valid today.

Rubies are defined as the red variety of the mineral corundum, and their name appropriately comes from a Latin word meaning "red." Corundum of any other color is called sapphire. There are a couple of exceptions. Purple star sapphire from India is sold as star ruby, and pink sapphire is often sold in Asia as ruby. No matter what it's called, pink and purple corundum should be priced much lower than red corundum (all other factors being equal).

Judging ruby color, however, is more complex than determining if a stone is red, pink, or purple. To describe the color factors that affect its value, we need to understand the concept of color. It can be divided into three main elements:

Hue — Refers to a pure color such as red, orange, or orangy red.

Lightness/Darkness (Tone) — Refers to the amount of color present. The lightest possible stone is colorless, the darkest is black. **Tone** is another word for lightness/darkness. It will be described in this book by the following terms:

very light
light
medium light
medium
medium dark
dark
very dark

Color Purity

Describes the degree to which the hue is masked by brown or gray. **Intensity, strength**, and **saturation** are other words for color purity. This book will describe it with terms used by the GIA (Gemological Institute of America):

vivid (most pure)
strong
moderately strong
very slightly brownish or grayish
slightly brownish or grayish
brownish or grayish (least pure)

Evaluating Ruby Color

Even though it's debatable as to what are the most valuable ruby hues and tones, gem dealers agree that pure, vivid colors are far more desirable than dull, muddy, brownish colors. To learn to judge **color purity**, look at the red objects around you and ask yourself which reds look the brightest and which look the most drab. Also go to a jewelry store and look at some rubies and red garnets side by side and try to determine which ones have the least amount of brown. Just being aware of color purity will increase your sensitivity to it. This in turn will help you choose a more desireable ruby.

Judging the **lightness or darkness** of faceted gemstones is difficult because they don't display a single, uniform tone. They have light and dark areas which become more apparent as you rock the stones in your hand. To judge the tone of a faceted ruby, answer the following questions:

- What is your first overall impression of the tone? Use one of the above lightness/darkness terms to describe the stone. If it is light or very light, the stone is either a pink sapphire or a lower-quality ruby, depending on the country where it's sold. For rubies, medium to dark tones are preferred by the trade.

- Do you see nearly colorless, washed-out areas in the ruby? This is an indication of poor color, poor cutting, or both.

- What percentage of the ruby looks black? If more than 90% of it is blackish, gem dealers would classify it as undesirable. The purpose of owning colored stones is to see color, not black.

Judging the **hue** of a ruby is just as hard as judging the tone. The different tones and possible brownish tints are distracting. Moreover, rubies are a blend of two colors--purplish red and orangy red. When you look at rubies from different directions while moving them, you can see these two colors. This is due to certain optical properties of corundum and cannot be removed.

When you judge the hue, look for the dominant color in the face-up view. What's your first overall impression? Generally, the more purple or orange a stone looks, the less it costs. The GIA, in their colored stone grading course, describes the most expensive ruby color as medium-dark, vivid red. Your chances of finding such a ruby are very slim. Normally, rubies have an orangy or purplish tint, and it's a matter of personal taste as to which is best. In Switzerland and France, orangy-red rubies are often preferred whereas in Japan, there is a tendency to prefer a slightly purplish red. Choose a color that you find attractive and that fits your budget, and follow the guidelines in the next section to make sure the stone looks as good at home as it did in the store.

How to Examine Color

The lighting and displays in jewelry stores are naturally designed to show gems at their best. To choose a ruby that will look good wherever you wear it and to detect value differences, follow these steps.

♦ First, clean the stone with a soft cloth if it's dirty. Dirt and fingerprints hide color and brilliance.

♦ Examine the stone face up against a variety of backgrounds. Look straight down at it over a non-reflective, white background and check if the center of the stone is pale and washed out. (This is undesirable). Then look at it against a black background. Do you still see glints of red or does most of the color disappear? Also, check how good the stone looks next to your skin.

♦ Examine the stone under direct light and away from it. Your ruby won't always be spotlighted as you wear it. Does it still show glints of red out of direct light? It should if it's of good quality.

♦ Look at the stone under various types of light available in the store. For example, check the color under an incandescent light-bulb, fluorescent light, and next to a window (The next section explains how the lighting affects color). If you're trying to match stones, it's particularly important to view them together under different lights. Stones that match under one light source may be mismatched under another.

♦ Every now and then, look away from the rubies at other colors and objects to give your eyes a rest. When you focus too long on one color, your perception of it is distorted.

♦ Examine the stone from the side to check if the color is evenly distributed. If the color is uneven or concentrated in one spot, this can sometimes decrease the stone's value.

♦ If you're looking for a ruby that is as red as possible, try finding a piece of paper, material, or an inexpensive synthetic stone that is a deep, vivid red. Use it as a basis for comparison when you shop. Keep in mind that colors seen in synthetic stones, paper, plastic, and fabrics may not be found in a natural gemstone. Nevertheless, comparison objects can help determine how purplish, orangy, or brownish a stone is, and using them is more reliable than trusting your color memory.

- Make sure you're alert and feel good when you examine stones. If you're tired, sick, or under the influence of alcohol or drugs, your perception of color will be impaired.

How Lighting Affects Ruby Color

Visualize how different the colors of a snow-capped mountain are at sunrise and midday. This difference is due to the lighting, not to a change in the mountain itself. Likewise, the color of a ruby will change depending on the lighting.

The whitest, most neutral light is at midday. Besides adding the least amount of color, this light makes it easier to see the various nuances of red. Consequently, it's best to judge ruby color under a daylight-equivalent light. Day-light fluorescent bulbs approximate this ideal, but some of these lights are better than others. One that is often recommended for colored-stone grading is the Duro-Test Vita Lite. Even though rubies are graded under daylight-equivalent light, they are generally displayed and look their best under incandescent light (light bulbs).

When you shop for rubies, your choice of lighting will probably be limited. Use the information below to help you compensate for improper lighting when you judge color.

Type of light	Effect of light
Light bulbs up to 150 watts and candlelight	Add red so rubies look redder.
Fluorescent lights	Depends on what type they are. Some have a neutral effect. Others add green, which can make red stones look grayish, orangy stones more orange, and purplish ones more purple.
Light under overcast sky or in the shade under blue sky	Adds blue so rubies look more purplish.
Sunlight	Depends on the time of day, the season of the year, and the geographic location. At midday, it normally has a neutral effect on the hue. Earlier and later in the day, it adds red, orange, or yellow; so purplish rubies will look redder then.

No matter what their hue, rubies will look brighter and less black in the more direct, intense light of mid-day, summer, or tropical sun. This difference in tone and color purity will also make them appear redder.

Another effect of sunlight is that its ultraviolet rays can cause a red fluorescence or glow in rubies, which also makes them look redder. Rubies from Burma are particularly noted for their strong red fluorescence.

Grading Color in Rubies Versus Diamonds

Grading color in rubies would be much easier if a scale of 23 letter grades could adequately describe their color differences. Diamond color is graded with a scale like this extending from D to Z. The jewelry trade, however, has not yet adopted a standardized system for grading colored stones. The following comparisons of diamond/ruby color grades and characteristics will help you understand why.

- Diamond color grades only need to indicate how light the stone is (its tone). Ruby color grades must also describe the hue and color purity to adequately explain price differences.

- Diamond color grades represent a smaller range of tones than is needed for rubies. The highest priced diamond tone, D, is colorless. Their lowest priced tone, Z, is light yellow. The tonal range of a ruby is still being debated and depends on the combination of the other two factors--hue and color purity. The trade agrees that rubies can be a medium to very dark red. Many Asian dealers feel the tone can extend to very light red (pink).

- Diamond color grades are based mainly on the side view of the stone. Ruby color grades are based mainly on the face-up view, which due to its many reflections is much harder to judge.

- Diamonds nearly always have one hue, if they are not colorless. Rubies are a blend of two hues, which complicates color grading. The cutting makes a difference in how these two hues combine in the face-up position.

- Diamonds can be color-graded against master stones. It would be too expensive and time-consuming to assemble master sets for rubies with all their variations of hue, tone, and color purity.

- The lack of color is what's important in diamonds (unless they're fancy-colored diamonds). The quality of the color is what's important in rubies, and the descriptive terms used must be applicable to all other colored gemstones for color comparison purposes. Naturally, a grading system that includes all colored stones will be far more complex than one just designed for diamonds.

Color Communication Devices

There are many color communication devices being used for gemstones. The three that are probably the most widely known are GemDialogue, the AGL (American Gemological Laboratories) system, and the GIA colored-stone grading system. All three share these two goals:

- To develop a practical description of color that can be used internationally when buying, selling, and appraising gems.

- To provide a set of standardized colors to which gemstones can be compared and matched.

A brief description of each system is given below.

GemDialogue

The GemDialogue system colors are offered as a master set of colors into which virtually any gem can fit. It has a color chart manual containing 21 spiral bound color charts on transparent acetate, each showing ten different saturation levels for each color (fig. 4.1). This makes it easy to compare the strong and weak colors which show up in a stone at the same time. If a stone is cut shallow, the strongest color will be on the rim of the stone and the weak color in the center. Both colors will always be on the same chart in such cases.

The degree to which colors are masked by brown or gray is determined by placing acetate overlays on the color charts. One of the overlays ranges from black to gray and the other one from brown to light brown. In all, the system gives you 6300 reference points in a very portable 4" x 8 1/2" manual. Also included are an instruction manual which explains the system in plain language and a grading and pricing manual which explains the pricing patterns of colored stones. The GemDialogue system is easy to learn and convenient to use when shopping for gems.

AGL (Color/Scan)

The AGL color grading system uses a set of color-comparison cards (Color/Scan) each of which has six oval holes (fig. 4.2). The holes are filled with layers of colored filters and a patterned foil that simulates the three-dimensional appearance of a gemstone color. The Color/Scan allows the user to view several samples simultaneously, like a set of diamond-color master stones. This ability enables the eye to utilize one of its most important features--to compare rather than remember color. The Color/Scan grading system is easy to use when shopping for gems and is explained in *The Color/Scan Training Manual.*

Unlike the other two systems, AGL's is being used on an internationally-recognized colored-stone grading document--the AGL colored-stone certificate (fig. 16.3, Chapter 16). Besides being an impartial report about gem quality and identity, this certificate helps Color/Scan users verify if they are grading stones according to AGL standards.

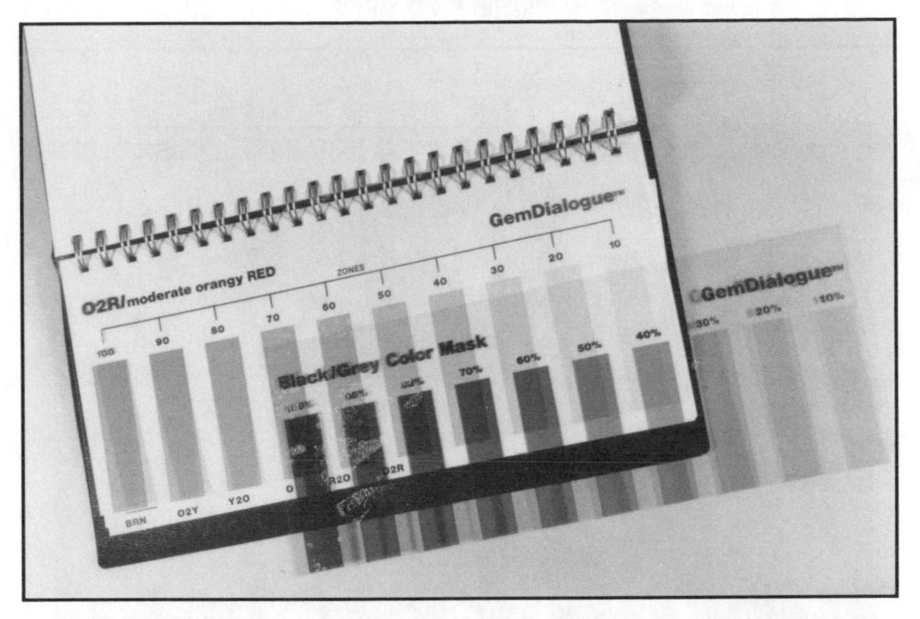

Fig. 4.1 GemDialogue color chart manual & black/grey color mask overlay

Fig. 4.2 Color/Scan grading cards (Photo courtesy AGL)

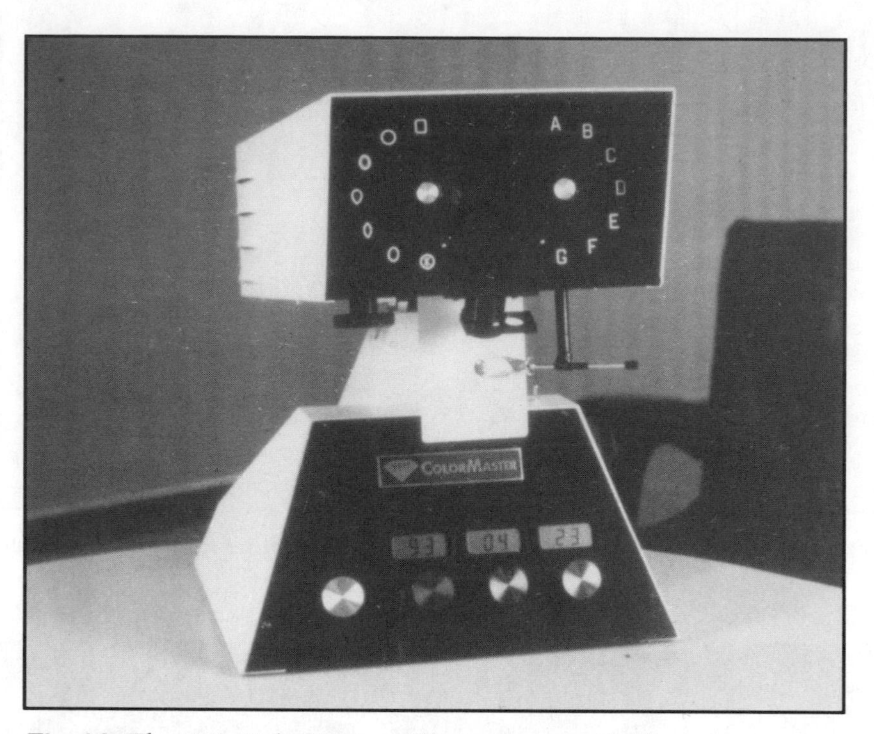

Fig. 4.3 The GIA ColorMaster (Photo Courtesy GIA)

GIA

The GIA colored-stone grading system uses as its color-comparison standard an instrument called the ColorMaster (fig. 4.3). It produces up to one million different color images of gems with a set of built-in filters that generate calibrated mixtures of red, green, and blue light. The ColorMaster, which is 19" x 15" x 12 3/4" (48 x 38 x 30 cm) and operates on either 110 volts-60Hz or 220 V-50Hz, is helpful as a laboratory tool for jewelers and appraisers.

It is easy for the layperson to understand GIA word descriptions of color even without any prior knowledge of the system (e.g. dark, strong orangy red). This color terminology is the most widely used in the trade largely because the GIA offers colored-stone grading seminars and home-study courses throughout the world and because it's easily understood by the public. Color terms such as "dark" and "strong" are defined and illustrated in these courses.

If you're interested in learning how these three systems evaluate gemstone color and how they can help you when buying and selling gems, it would be best to get information directly from the suppliers of each system. You can do this by writing to the following addresses:

GemDialogue Systems Inc.
205-02 26th Ave.
Bayside, NY 11360
(718) 352-0009

American Gemological Laboratories
580 Fifth Ave. Suite 1211
New York, NY 10036
(212) 704-0727

Gemological Institute of America
1660 Stewart Street
Santa Monica, Calif. 90404
(800) 421-7250 (within the USA)
(213) 829-2991

Chapter 4 Quiz

Select the correct answer.

1. Under what type of light will a ruby look reddest?

   a. A fluorescent light
   b. A standard 100-watt light bulb
   c. Daylight under a cloudy sky
   d. A daylight-equivalent lamp

2. Which of the following terms refers to how light or how dark a gemstone is?

   a. Hue
   b. Tint
   c. Tone
   d. Brightness

3. Which of the following can affect your perception of color?

   a. Lighting
   b. Medications
   c. The color of your clothes
   d. All of the above

4. It's easier to grade the color of diamonds than of rubies because:

   a. Diamonds are colorless.
   b. Diamond color grades only indicate how light or dark the color is.
   c. Diamonds don't have as many rainbow-like interference colors as rubies.
   d. Diamonds are controlled by a cartel which has standardized how they are graded.

5. Beryl bought a ruby ring in Bangkok at Christmas. When she got back to her home in Canada, the ruby didn't look as red and bright as it did in Bangkok. Explain why.

   a. The jeweler probably switched the stone.
   b. Canadians have a distorted sense of color.
   c. The sunlight in Canada doesn't bring out the red of a ruby as well as the tropical sunlight of Bangkok.
   d. Beryl probably drank too much alcohol while she was in Bangkok.

True or False

6. Grayish red is a less valuable color for a ruby than a pure slightly purplish red.

7. It's best to examine rubies against a red background.

8. The darker the ruby, the more valuable it is.

9. It's easier for people to compare color than to remember it.

10. In some countries pink and purple sapphires are called rubies.

11. A standardized system for grading the color of rubies has not been adopted yet by the jewelry trade.

12. The best way to judge the color of a ruby is to look at it for a long time so your eyes and mind can absorb as much of the color as possible.

Answers:

1. B
2. C
3. D
4. B
5. C.
6. True
7. False
8. False - Not always. Very dark rubies with a lot of black areas are not highly prized.
9. True
10. True
11. True
12. False - Focusing on one color for a long period can distort your perception of it. It's best to glance at other colors every now and then.

5

Judging Sapphire Colors

Sapphire means blue and it probably comes from a Greek word meaning blue; yet sapphires are not necessarily blue. Perhaps you're wondering why.

It was just a couple hundred years ago that scientists discovered blue sapphire was chemically the same mineral as "Oriental amethyst," "Oriental emerald," "Oriental topaz," and ruby. Afterwards, the term "sapphire" was expanded to include all varieties of the mineral corundum except ruby. Consequently, "Oriental emerald," for example, is now called "green sapphire."

Before 1980, rarely did jewelers sell sapphires that were not blue. Part of the reason was that a large percentage of **fancy sapphires** (non-blue sapphires) were pale and unattractive. Around 1980, some dealers in Thailand started to intensify the color of fancy sapphires with controlled heat treatments (see Chapter 11). Intense yellow sapphires, in particular, became plentiful. Also, new sources of fancy sapphire were found in countries like Tanzania. So today, you can find pink, orange, purple, yellow, or green sapphires in jewelry stores. You may have to go to a museum, though, to see fancy sapphires of exceptional quality.

Blue Sapphires

When used by itself, the term **sapphire** still normally refers to the blue variety. In its highest qualities, it is more expensive than the other sapphire colors. Top-quality Kashmir sapphires, for example, can cost over $15,000 a carat.

Frequently, the best color of sapphire is described as a cornflower blue. Most likely, your mental image of this color is different than that of your jeweler. Even though cornflowers grow like weeds in Europe and northern Asia, many people in North America and in the Southern hemisphere have never seen a blue cornflower. Cornflowers come in varying shades of blue and violet as well as pink, purple, white, and yellow, so "cornflower blue" evokes a wide array of color images even to people that have seen the flower. Since terms like "cornflower blue," "inky blue," and "pigeon-blood red" are very ambiguous, they will not be used in this book to describe rubies and sapphires.

There are differences of opinion as to what is the best sapphire hue. Some say blue; others say violetish blue. Most dealers agree, however, that greenish blues are less valuable. Dealers also have different tone preferences. Some prefer medium tones of blue. Others prefer medium-dark tones. The GIA in its colored-stone grading course (1989 charts) describes the most expensive sapphire color as either

medium-dark, vivid blue
or medium-dark, vivid violetish blue.

If you don't plan to resell your stone, there's no need for you to base your choice of hue on trade preferences. If you prefer slightly greenish blues, and they look good on you, take advantage of their lower price. Actually, it's normal for blue sapphires to have some green. Just as rubies are a blend of two colors, so are sapphires. Their blue usually results from a mix of greenish blue and violetish blue.

When judging the color of a sapphire, ask yourself the following questions:

♦ Does the sapphire look grayish? The more gray that is present, the lower the value. Vivid colors are most desirable.

♦ What percentage of the stone looks black? It's not uncommon for over 90% of a sapphire to appear black. This greatly reduces its price because the sapphire is valued for its blue color.

♦ Does the center of the stone or overall color look pale and washed out? This lowers the value. The most preferred tones for sapphire range from medium light to medium dark.

♦ Is the color evenly distributed in the stone? Sapphires are more likely than rubies to have patches or bands of differing colors or tones. Some people like this patterned effect even in the face up view of a stone, but the trade generally places a higher value on sapphires with a uniform color.

Figure 5.1 is an example of a sapphire with layers of varying color. It's crown is almost colorless. Since face up its color looks even, this stone would normally be considered acceptable. However, if it had a uniform color throughout, there would be a tendency to consider it more valuable.

♦ Under what kind of lighting will you normally wear the sapphire? Try to view it under that lighting to be sure you like the color. As is the case with rubies, it's a good idea to view sapphires under various light sources and against white and black backgrounds as well as your skin.

Fig. 5.1 A sapphire with layers of color

Fancy Sapphire

Padparadscha

Padparadscha, a pinkish-orange stone found in Sri Lanka, is the rarest and most prized of all the fancy sapphires. Its name is believed to have come from the Sinhalese word for the lotus flower, which has a similar color. Frequently, orange sapphire is called padparadscha, but most corundum dealers agree that both pink and orange hues must be present for a stone to be a true padparadscha.

This stone can range from a light to medium tone and from a pinkish orange to orange-pink hue. If the colors look a bit brownish, the value is greatly reduced, and the stone may lose its classification as a padparadscha. As with all sapphires, vivid colors are the most prized.

Due to their rarity, you should not expect your local jewelers to have padparadschas in stock. They could even have a hard time finding one for you, but if they do find you one, expect to pay a high price. In their finest qualities, true padparadschas in large sizes can cost over $10,000 per carat.

Pink Sapphire

Next to the padparadscha, pink sapphire is the most highly prized of all the fancy sapphires. Since "pink" is a synonym of "light red" and since fine rubies cost more than sapphires, many Asian dealers prefer to call pink sapphires rubies. Historically, though, Asian cultures have differentiated between the two by referring to pink sapphires as either "female" or "unripe" rubies.

The jewelry trade in western countries prefers to treat the pink sapphire as a unique stone with its own merits, rather than as a second-rate ruby. High-quality pink sapphires are more rare than rubies or blue sapphires and can cost several thousand dollars per carat. In their lowest qualities, however, the price of pink sapphires can fall below $100 per carat, even for stones as large as 3 carats.

The most valuable pink sapphires (according to the GIA's colored-stone grading charts) range in hue from vivid purple-red to reddish purple and have a medium-light tone. As the stones get lighter, more brownish, or more purple, their value decreases.

Orange Sapphire

During the 1980's there was a significant increase in the production of orange sapphire, thanks to expanded mining in Tanzania's Umba River Valley. The orange sapphire, often mistakenly called "padparadscha," ranges from a yellowish-orange hue to an orangy red. Vivid, red-orange stones with medium-dark tones are the most valued, but they generally sell for about 1/3 to 1/2 the cost of a pink-orange padparadscha of similar size and quality. Considering the high status the color red holds in the gemstone world, it's curious that the pink-orange padparadscha is worth more than a red-orange sapphire. Supply, demand, and tradition, more than logic, is what sets prices in the jewelry trade.

Purple & Color-change Sapphire

Purple sapphire is a step down in price from orange sapphire. The more red and less brown or gray a purple sapphire has, the greater its value. Medium-dark, purple-red stones, which are sometimes called **plum sapphires** or **amethystine sapphires**, are the most prized.

Some sapphires that look purple or violet indoors under an incandescent light-bulb look blue to grayish-blue when viewed in daylight or under a fluorescent light. In rare cases, sapphires that are green outdoors turn reddish brown indoors. The stronger the color change and the purer the colors, the more of a collector's item the sapphires are. In the trade, these stones are appropriately called **color-change** or **alexandrite-like sapphires.**

Yellow Sapphire

Compared to the preceding fancy stones, yellow sapphires are fairly common. Even if jewelers don't have them in stock, they can get them for you. Yellow sapphires range in color from greenish yellow to orangy yellow. Strong light yellows or orangy yellows are the most valued. The least valued are the very pale or brownish stones.

Due to their light color and high transparency, flaws are more visible in yellow sapphires than in the other color varieties. Consequently, it's more important for yellow sapphires to have a good clarity. Prices for these stones are similar to those of purple sapphires. You should be able to find, for example, a high-quality yellow sapphire from 1 to 3 carats, for less than $1000 per carat.

Green Sapphire

Green sapphire is the lowest-priced transparent gem-quality sapphire. One of the reasons its price normally falls below $200 a carat is that it does not come in strong or vivid colors. Another reason is that green sapphire is often just lower-quality dark-blue sapphire that is cut at a different angle to show its greenish blue color. (As mentioned earlier, the blue in sapphires is a blend of violetish blue and greenish blue).

Green sapphire ranges from a blue-green to yellow-green. Its most valuable colors are a slightly or very slightly bluish green of a medium-dark tone. If pure green sapphires with no gray or brown are ever found, they will sell for a lot more.
Stones with emerald-like colors, such as jade and garnet, can command very high prices.

What Color is Best for You?

To answer this question, consider the following factors:

- **Your purpose for buying the stone**. If you're buying a stone just for personal pleasure, it doesn't matter which color you choose as long as it looks good on you. But if you plan to resell it later, you would probably be better off to avoid grayish and brownish colors and stones with very light or dark tones.

- **Your budget**. Color has a major affect on the price of a stone. A medium or medium-dark, pure blue sapphire may not fit your budget, but a slightly different color could be in your price-range.

- **The availability of the color**. If you're looking for a sapphire with a very pure blue color and money is no object, you still may have a hard time finding it in the size and quality you want. If you're serious about the stone, your jeweler can call around to various dealers to try to find you one. Auctions can also be a good source for rare stones. In the end, though, you may have to compromise on either the color, quality, or size.

- **The clarity of the stone**. Darker colors mask flaws much better than lighter colors. If you prefer light-colored stones, make sure their flaws are not so obvious that they draw attention away from the color and overall beauty of the stone. Otherwise, you might be better off choosing a darker color.

- **Your wardrobe.** If you buy jewelry that blends with your existing wardrobe, you will save money by not having to buy new clothes, and you will be able to wear the jewelry more often.

- **Your complexion and hair color**. Carole Jackson in her book *Color Me Beautiful* shows us how our hair and skin coloring can determine what colors look best on us. On page 28, she points out that most blacks and Orientals and people with pinkish, rosy, or olive skin look best in clear, vivid, cool colors, including purplish and pinkish reds. On the other hand, warm colors like orangy red are flattering to people with peachy or golden complexions (Redheads and blondes often have this skin coloring).

Knowing how orangy red can either compliment or detract from a person's appearance helps us understand why in parts of Europe orangy-red rubies are often preferred to those that are purplish red whereas in Japan there tends to be a preference for purplish-red rubies over orangy rubies. It's natural for people to be attracted to the colors that look best on them.

Jorge Miguel in his book *Jewelry How to Create Your Image* (Chapter 3) shows us why women should consider their hair color when choosing between light- or dark-toned gemstones for necklaces and earrings. On blondes and redheads, darker stones can stand out and attract more attention than pastel colors. On the other hand, pastel stones can provide a more striking contrast than dark stones on brunettes or women with black hair. Medium and medium-dark stones can look good on women of any hair color.

♦ **Your personal preference and personality.** This should play a major role in your choice. Instinctively, you often prefer what looks best on you. According to color psychologists, there can also be a link between your personality and your color preferences. To see what these links may be, find your favorite color in the list below, and check if the traits commonly associated with that color describe your personality.

Favorite Color	Typical Personality Traits
Blue	Conscientious, sensitive, stable, cautious, responsible, introspective.
Green	Emotionally well-balanced, reliable, calm, status-conscious, frank, civic-minded.
Yellow	Intellectual, assertive, cheerful, logical, idealistic, introspective.
Orange	Sociable, cheerful, agreeable, good-natured, extroverted.
Red	Passionate, bold, aggressive, dynamic, impulsive, extroverted.
Purple & Violet	Artistic, adaptable, observant, intelligent, witty, introspective.

Sometimes, we expect jewelry salespeople to tell us what color is best for us. They can guide us in our choices, but in the end, we need to consider all of the above factors and then make the final decision ourselves when buying gems for our personal pleasure. If we do, we should end up with stones that will not only enhance our appearance but that can also bring us long-term enjoyment.

6

Carat Weight

The term "carat" originated in ancient times when gemstones were weighed against the carob bean. Each bean weighed about one carat. Gem traders were aware, though, that the weights varied slightly. This made it advantageous for them to own both "buying" beans and "selling" beans.

In 1913, carat weight was standardized internationally and adapted to the metric system, with one carat equalling 1/5 of a gram. The term "carat" sounds more impressive and is easier to use than fractions of grams. Consequently, it is the preferred unit of weight for gemstones.

The weight of small stones is frequently expressed in **points**, with one point equaling 0.01 carats. For example, five points is the same as five one-hundredths of a carat. Contrary to what is sometimes assumed, jewelers do not use "point" to refer to the number of facets on a stone. The following chart gives examples of written and spoken forms of carat weight:

Written	Spoken
0.005 ct (0.5 pt)	half point
0.05 ct	five points
0.25 ct	twenty-five points or quarter carat
0.50 ct	fifty points or half carat
1.82 cts	one point eight two (carats) or one eighty-two

Note that "point" when used in expressing weights over one carat refers to the decimal point, not a unit of measure. Also note that "pt" can be used instead of "ct" to make people think for example, that a stone weighs 1/2 carat instead 1/2 of a point.

Effect of Carat Weight on Price

Most people are familiar with the principle, the higher the carat weight the greater the value. However, in actual practice, this principle is more complicated than it appears. This can be illustrated by having you arrange the following four ruby rings in the order of decreasing value. Assume that the quality and shape of all the rubies are the same and that the ring mountings have equivalent values.

a. 1-carat ruby solitaire ring
b. cocktail ring, 12 rubies, 1 carat TW
c. 2-carat ruby solitaire ring
d. cocktail ring, 24 rubies, 2 carat TW

In almost all cases, the order of decreasing value would be c>a>d>b. Strangely enough, a single one-carat ruby usually costs more than two carats of small rubies of the same quality. This is because the supply of large rubies is very limited. So when you compare jewelry prices, you should pay attention to individual stone weights and **notice the difference between** the labels **1 ct TW** (one carat total weight) **and 1 ct** (the weight of one stone). A jewelry piece with a **1 ct** top quality ruby or sapphire can be worth more than 10 times as much as a piece with **1 ct TW** of stones of the same quality.

When comparing the cost of rubies and sapphires, you should also start noting the **per carat cost** instead of concentrating on the total cost of the stone. This makes it easier to compare prices more accurately, which is why dealers buy and sell gems using per-carat prices. The following equations will help you calculate the per-carat cost and total cost of rubies and sapphires.

$$\textbf{Per-carat cost} = \frac{\textbf{stone cost}}{\textbf{carat weight}}$$

Total cost of a stone = carat weight x per-carat cost

The per-carat prices of rubies and sapphires are listed in terms of either their weight or millimeter size (unlike those of diamonds, which are usually only listed according to carat weight). Corundum stones over 1/2 to 3/4 of a carat are generally priced according to weight, whereas those under 1/2 carat tend to be listed in terms of millimeter size. Listed below are some price/weight categories for rubies and sapphires. Colored-stone weight categories tend to be broad and flexible.

0.50 or 0.75 to .99 ct (this category varies from dealer to dealer)
1.00 - 1.49 cts
1.50 - 1.99 cts
2.00 - 2.99 cts
3.00 - 3.99 cts

As mentioned before, rubies and sapphires weighing less than 1/2 carat are often assigned per carat prices according to millimeter size. There's no point in learning size categories since they vary from one dealer to another. Just be aware that shape and carat weight can affect the per carat value of rubies and sapphires and follow these two guidelines:

♦ Compare per carat prices instead of the total cost.

♦ When judging prices, compare stones of the same size, shape, quality, and color.

Size Versus Carat Weight

Sometimes in the jewelry trade, the term "size" is used as a synonym for "carat weight." This is because size and weight are directly related. However, as corundum stones increase in weight, their size becomes less predictable. This means that a 0.90 carat sapphire may look bigger than a 1.05 carat sapphire. Therefore, you need to consider stone measurements as well as carat weight when buying rubies and sapphires. You don't need to carry a millimeter gauge with you when you go shopping. Just start noting the different illusions of size that various stone shapes and measurements can create.

You should also note that rubies and sapphires normally have different measurements than other gemstones of the same weight. For example, because of its higher density, a one-carat ruby is considerably smaller than a one-carat emerald. Rubies and sapphires are heavier (more dense) than diamonds and most colored stones. One major exception is the garnet. Depending on the type, garnets may have a similar or greater density than corundum stones.

Estimating Carat Weight

If you buy jewelry in a reputable jewelry store, you normally don't need to know how to estimate the carat weight of gems because the weight will be marked. However, if you buy jewelry at flea markets, garage sales, or auctions, it is to your advantage to know how to estimate weight.

One way to estimate the weight of faceted rubies and sapphires is to measure their length and width (or diameter) with a millimeter gauge (these are sold at jewelry supply stores). Then match the measurements to those of table 7.3, and look at the corresponding weights. This works best with stones that are small, well-cut, and calibrated (cut to specific sizes). This is not a good way, however, of estimating the weight of stones that have deep bulging pavilions, flat profiles, or odd measurements. It's better to measure their depth as well as the length & width, and then calculate the weight using tables 7.4 to 7.6. Of course, the only accurate means of determining the weight of a stone is to take it out of its setting and weigh it. This, however, is not always possible nor advisable.

Table 7.1 Weight Conversions	
1 carat (ct)	= 0.2 g = 0.006 oz t = 0.007 oz av = 0.31 dwt
1 gram (g)	= 5 cts = 0.032 oz t = 0.035 oz av = 0.643 dwt
1 ounce avoirdupois (oz av)	= 28.3495 g = 0.911 oz t = 18.229 dwt = 141.75 cts
1 troy ounce (oz t)	= 31.103 g = 1.097 oz av = 20 dwt = 155.51 cts
1 pennyweight (dwt)	= 1.555 g = 0.05 oz t = 0.055 oz av = 7.776 cts

Table 7.2 Millimeter Sizes (Courtesy Grieger's Inc.)

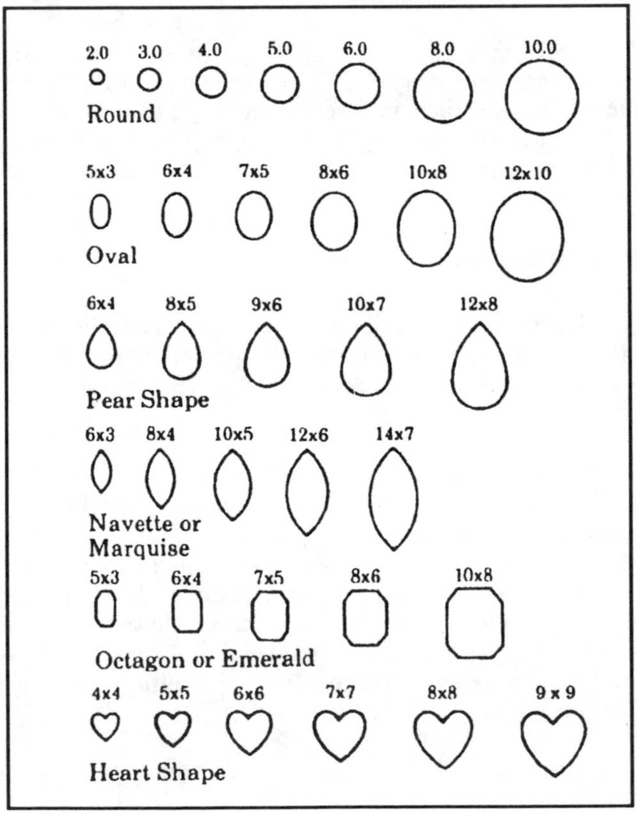

2.0 3.0 4.0 5.0 6.0 8.0 10.0
Round

5x3 6x4 7x5 8x6 10x8 12x10
Oval

6x4 8x5 9x6 10x7 12x8
Pear Shape

6x3 8x4 10x5 12x6 14x7
Navette or
Marquise

5x3 6x4 7x5 8x6 10x8
Octagon or Emerald

4x4 5x5 6x6 7x7 8x8 9 x 9
Heart Shape

Table 7.3 Sizes & Approximate Weights of Calibrated, Faceted Corundum

SHAPE	SIZE	APPROX WT	SHAPE	SIZE	APPROX WT
Round	2.0 MM	0.04- 0.05	Emerald Cut	5x3 MM	0.30- 0.36
	2.5	0.08- 0.09		6x4	0.57- 0.70
	3.0	0.13- 0.15		6.5x4.5	0.78- 0.97
	3.5	0.20- 0.23		7x5	1.00- 1.30
	4.0	0.29- 0.35		7.5x5.5	1.35- 1.65
	4.5	0.39- 0.50		8x6	1.80- 2.10
	5.0	0.53- 0.65		9x7	2.90- 3.15
	5.5	0.71- 0.90			
	6.0	0.93- 1.15	Princess Cut	2.00	0.07
	6.5	1.25- 1.50	(Square)	2.25	0.11
	7.0	1.50- 1.85		2.50	0.13
	7.5	1.80- 2.25		3.00	0.22
	8.0	2.35- 2.85		3.50	0.33
Oval	5x3	0.25- 0.33	Triangle	3.00	0.14
	6x4	0.48- 0.58		3.50	0.22
	6.5x4.5	0.65- 0.79		4.00	0.33
	7x5	0.85- 1.05		4.50	0.48
	8x6	1.45- 1.70		5.00	0.60
	9x7	2.15- 2.65		5.50	0.83
	10x8	3.15- 3.75		6.00	1.23
Pear	5x3	0.21- 0.27	Marquise	6x3	0.23- 0.29
	6x4	0.40- 0.50		8x4	0.54- 0.70
	7x5	0.73- 0.90		9x4.5	0.80- 0.98
	8x5	0.85- 1.00		10x5	1.06- 1.35
	9x6	1.45- 1.70		12x6	1.80- 2.40
	10x7	2.00- 2.55			

Note: These weights are only guides. The actual weights can vary. The information in this table is based mostly on size/weight/shape lists of Overland Gems, Inc. and Chatham Created Gems.

Table 7.4 Weight Estimation Formulas for Faceted Rubies & Sapphires

Shape	Formula
Rounds	$Diameter^2$ x depth x 4.00 x .0018
Ovals	$Diameter^2$ x depth x 4.00 x .0020 (Average out length and width to get diameter)
Square Cushion	$Diameter^2$ x depth x 4.00 x .0018 (Average out horizontal, vertical, and diagonal measurements for diameter)
Rectangular Cushion	$Diameter^2$ x depth x 4.00 x .0022 (Average length and width to get diameter)
Square Emerald Cut	Average $width^2$ x depth x 4.00 x .0023
Rectangular Emerald Cut	Length x width x depth x 4.00 x .0025
Square (with corners)	Average $width^2$ x depth x 4.00 x .0024
Rectangular Baguette	Length x width x depth x 4.00 x .0026
Pear	Length x width x depth x 4.00 x .0018
Marquise	Length x width x depth x 4.00 x .0017
Heart	Length x width x depth x 4.00 x .0021
Cabochons	Length x width x depth x 4.00 x factor listed in Tables 7.5 and 7.6

Note: These formulas are based on stones with medium girdles, no pavilion bulge, and well-proportioned shapes. Thick girdles may require a correction of up to 10%. Bulging pavilions may require a correction as high as 18%. The correction for a poor shape outline can be up to 10%. To estimate the weight of other colored stones, substitute the specific gravity of corundum, 4.00, with their specific gravity.

The above information is based on handouts from the GIA appraisal seminar.

	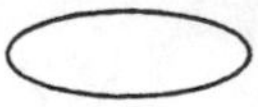		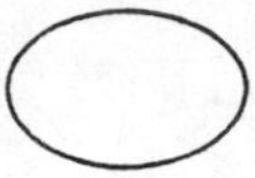	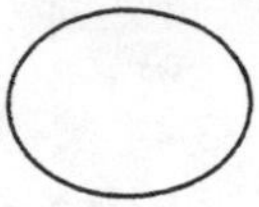	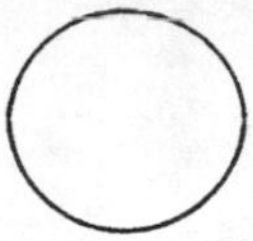
	.0029	.0029	.0029	.0029	.0029
	.0025	.0025	.0023	.0024	.0024
	.0026	.0026	.0024	.0025	.0025
	.0027	.0027	.0025	.0026	.0026
	.0028	.0028	.0026	.0027	.0027
	.0029	.0029	.0027	.0028	.0028
	.0029	.0029	.0029	.0029	.0029
	.0026	.0026	.0026	.0026	.0027
	.0027	.0027	.0027	.0027	.0028
	.0028	.0028	.0028	.0028	.0029
	.0029	.0029	.0029	.0029	.0030
	.0023	.0023	.0023	.0023	.0023
	.0023	.0023	.0023	.0024	.0024

Table 7.5 Factor chart for cabochon weight estimation--Length x width x depth x 4.00 x factor (Courtesy Gemological Institute of America)

.0023	.0023	.0024	.0025	.0025
.0023	.0024	.0025	.0026	.0026
.0024	.0025	.0026	.0027	.0028
.0024	.0024	.0024	.0026	.0026
.0026	.0026	.0026	.0028	.0028
.0028	.0028	.0028	.0030	.0030
.0026	.0026	.0026	.0026	.0026
.0028	.0028	.0028	.0028	.0028
.0030	.0030	.0030	.0030	.0030
.0024	.0024	.0024	.0024	.0024
.0026	.0026	.0026	.0026	.0026
.0028	.0028	.0028	.0028	.0028

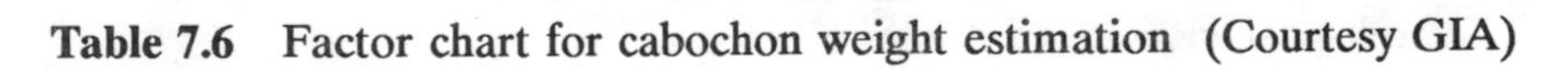

Table 7.6 Factor chart for cabochon weight estimation (Courtesy GIA)

Quiz (Chapters 5 and 6)

True or False?

1. Purple is a more highly valued sapphire color than pink.

2. Sapphires with bands or patterns of different color normally cost more than those with one even color.

3. A 2-carat round ruby is larger than a 2-carat round diamond.

4. The less grayish a sapphire is, the greater its value.

5. The carat is a unit of weight equalling 0.20 gram.

6. Orange sapphires usually sell for more than yellow sapphires.

7. Appraisers can determine the exact weight of a ruby by measuring it and calculating the weight with the aid of mathematical formulas.

8. Greenish blue is normally considered a less valuable color for a sapphire than violetish blue.

9. When estimating the weight of rubies and sapphires, you should take into account their shape and proportioning.

10. The darker a sapphire is, the greater its value.

11. A 1.5-carat, good-quality ruby is worth less than 10 rubies of the same color and quality which have a total weight of 2.5 carats.

Answer the following questions:

12. What two colors do true padparadschas have?

13. What are non-blue sapphires called?

14. If a sapphire weighs 3 carats and costs $3900, what is its per-carat price?

15. What's the total cost of a 1/4 carat ruby that sells for $240 a carat?

16. What would be the estimated weight of a well-proportioned pear-shape sapphire that is 8.00 mm long, 5.55 mm wide and 3.70 mm deep?

Answers:

1. False

2. False

3. False - Sapphires are heavier and have a higher density than diamonds. Consequently, the 2-carat ruby would be smaller than the 2-carat diamond.

4. True

5. True

6. True

7. False - The only way to determine the **exact** weight of a stone is to weigh it.

8. True

9. True

10. False - Extremely dark sapphires with a lot of black areas are not highly valued.

11. False - Good-quality rubies over 1 carat have a much greater per-carat value than small rubies of the same quality.

12. Pink and orange

13. Fancy sapphires or fancy-color sapphires

14. \$1300 $\frac{\$3900}{3} = \$1300$

15. \$60 1/4 x \$240 = \$60

16. 1.18 carats 8.00 x 5.55 x 3.70 x 4.00 x .0018

7

Judging Clarity

Imagine looking at the stars through a telescope and seeing comets, galaxies, falling stars, satellites, Saturn and other planets. That's a bit what it's like to look through a microscope at rubies and sapphires.

When you try to find forms within these stones or marks on their exterior, you are analyzing their **clarity**. Clarity is the degree to which a stone is free from external marks called **blemishes** and internal features called **inclusions**. Sometimes the jewelry trade refers to them as **clarity characteristics** or **identifying features**. "**Flaw**" is the term that is normally used in this book since it is shorter and clearer. It refers to both blemishes and inclusions.

Even though terms like "flaw" and "blemish" have negative connotations, their presence can be positive. Flaws are identifying marks that can help you identify your stone at any time. They can lower the price of a stone without affecting its beauty. They can also increase the value of a stone by helping prove that it's from somewhere like Burma or Kashmir. (These places have a reputation for producing top-quality stones.) Flaws are especially important as evidence that your stone is natural. Jewelers and dealers are suspicious of flawless rubies and sapphires because that's usually a sign that the stone is synthetic (man-made). Therefore, instead of looking for a flawless stone, try to find one whose beauty and durability is not affected by its blemishes and inclusions. This chapter will help you do this, but first you should know what flaws are normally found in rubies and sapphires.

Inclusions in Natural Ruby & Sapphire

♦ **Cracks** of various sizes (figs. 7.1 & 7.2) are commonly seen in corundum (ruby & sapphire). When cracks are jagged, they're called **fractures**. Cracks in corundum are called **parting** if they're straight and flat and lie along **twinning planes**. These twinning planes are the result of alternating layers of corundum growing in different directions. Because of their appearance, cracks are often referred to as **feathers** in the trade.

♦ **Crystals** are solid mineral inclusions of various shapes and sizes (figs 7.3 & 7.6). Examples of minerals found in corundum are pyrite, garnet, zircon, calcite, and spinel. Minute crystals are sometimes called **pinpoints** or **grains,** and when they are grouped together, they may look like comets, galaxies, or falling stars.

♦ **Negative crystals or voids** (fig. 7.3) are hollow spaces inside a stone that have the shape of a crystal. They often resemble solid crystals, so for purposes of clarity grading, they're just called "included crystals."

♦ **Needles** are long, thin inclusions that are either solid crystals or tubes filled with gas or liquid (fig. 7.4).

♦ **Silk** in corundum consists of very fine fibers of rutile (titanium dioxide) or other minerals (fig. 7.5). It can also be made of mineral grains arranged in straight rows. These fibers or rows intersect and resemble silk, hence the name. Well-formed silk can be proof that a stone was not heat treated to improve its color. Very high temperatures tend to dissolve it and make it look fuzzy or dot-like. Since untreated stones are more valued than treated ones, the presence of clear, well-formed silk can be a welcome sign.

♦ **Fingerprints**, common in corundum, are partially healed cracks (figs. 7.3 & 7.9). Rubies and sapphires grow from a mineral solution, and if they split during formation, the solution can fill the cracks and let them grow back together. During this healing process, stray drops of liquid are sealed in and form patterns that look like human fingerprints.

♦ **Halos** are circular fractures surrounding a crystal (fig 7.6). These structures, which often resemble Saturn with its rings, generally result from tension created by the growth of the crystal inside the halo or by heat treatment.

♦ **Growth or color zoning** refers to an uneven distribution of color in a stone (fig. 7.7). If the different color zones look like bands, they are called **growth or color bands** (fig. 7.8). Straight or hexagonal banding often occurs in sapphires and sometimes occurs in rubies.

♦ **Cavities** (fig. 7.10) are holes or indentations extending into a stone from the surface. Cavities can result when solid crystals are pulled out of a stone or when negative crystals are exposed during the cutting process.

♦ **Chips** are notches or broken off pieces of stone along the girdle edge or at the culet.

Fig. 7.1 A serious crack (feather) across a ruby.

Fig. 7.2 A minor crack

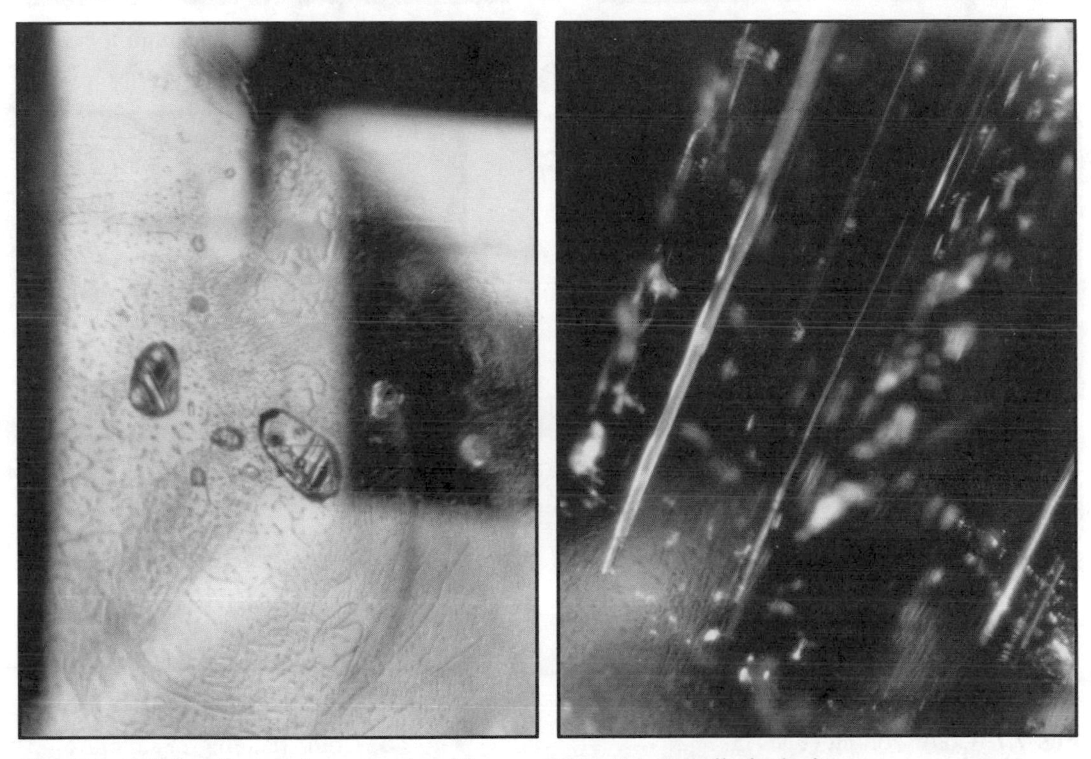

Fig. 7.3 Solid and negative crystals with a fingerprint inclusion in the background

Fig. 7.4 Needle inclusions

Fig. 7.5 Silk inclusions

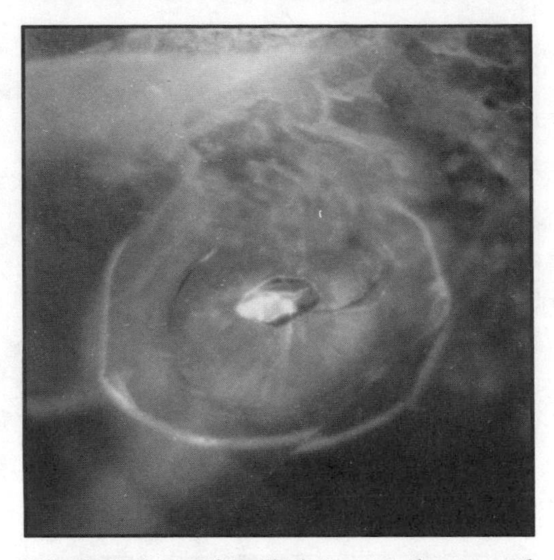

Fig. 7.6 A tension "halo" around a crystal

Fig. 7.7 Color zoning (uneven coloring) in sapphire (pavilion view)

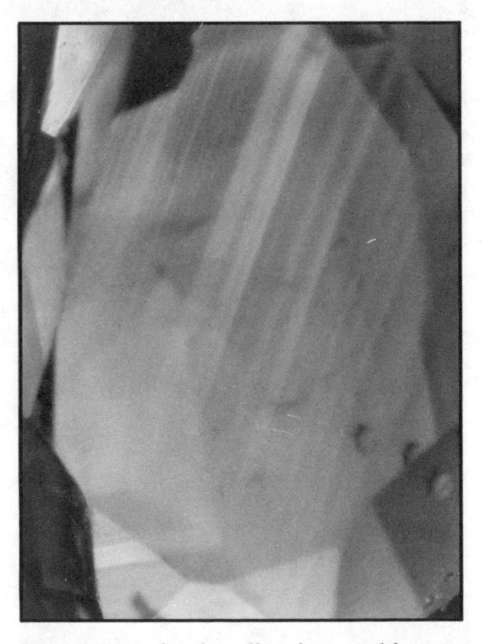

Fig. 7.8 Color banding in sapphire

Surface Blemishes on Rubies & Sapphires

- **Scratches** are straight or crooked lines scraped on a stone. Since they can be polished away, they don't have much of an effect on the clarity grade.
- **Pits** are tiny holes on the surface of a stone that often look like white dots (fig. 7.11).
- **Nicks** are flaws along the edge of a girdle or facet where bits of stone have broken away.
- **Abrasions** are rough, scraped areas usually along the facet edges of a stone (fig. 7.11). Abrasions are seen more often on colored stones than on a diamond, due to the diamond's exceptional hardness.

Colored Stone Clarity Grading Systems

A variety of clarity grading systems are being used for judging and appraising colored stones. Two that are sometimes used are the ones developed by the GIA and the AGL (American Gemological Laboratories).

The GIA and AGL systems differ in the following ways:

- AGL clarity grading is done with the unaided eye. GIA clarity grading is done with both the unaided eye and 10-power magnification.
- AGL uses different grade names for colored stones than it does for diamonds, and it uses comments to clarify the meanings of these grades. The GIA uses diamond grade names like "VS" and "SI" for colored stones but assigns these names a different meaning from diamond. For example, VS diamonds generally have no flaws that are visible to the unaided eye, but VS rubies often have them.
- AGL grades all colored stones with one set of grades which each have one meaning.
  The GIA divides colored stones into three clarity types:
  1. Stones like aquamarine which are are expected to be relatively free of inclusions.
  2. Stones like rubies which are expected to contain minor inclusions.
  3. Stones like emeralds which are expected to have many visible inclusions.

  Since the average clarity of the different gem types varies, the GIA feels it would be unfair to grade them the same. To eliminate unequal comparisons, the GIA uses a different set of grading definitions for each of the three clarity types. The grade names, however, remain the same. Consequently, a grade like "VS" has three possible meanings when applied to colored stones.

Both the GIA and AGL systems are helpful for classifying the clarity of colored stones. However, from the standpoint of the layperson, the AGL system is easier to use and understand. To grade colored stones with the AGL system, you only need to learn one set of grades and definitions. When jewelers use GIA clarity grades for different gems, keep in mind that grades like VS have four meanings, three for colored stones and one for diamonds.

Fig. 7.9 Fingerprint inclusions

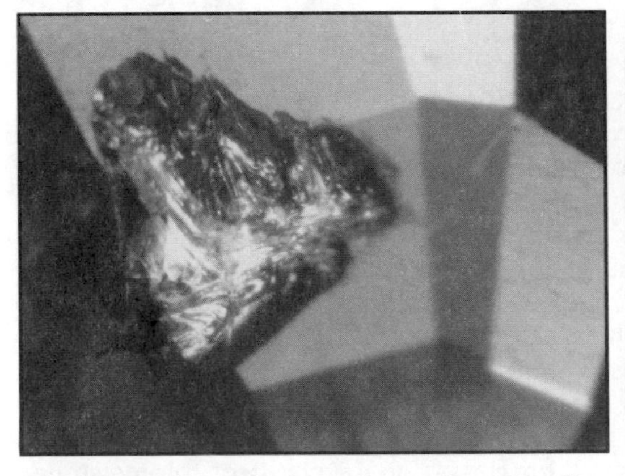

Fig. 7.10 Close-up view of a cavity. See fig. 8.8 in Chapter 8 for broader view of stone

Fig. 7.11 Abrasions and pits on the surface of a sapphire

Since both the GIA & AGL systems are currently being used in the jewelry trade, you should be familiar with each of them. This will help you understand appraisals and gem-lab reports. The GIA and AGL clarity grades for rubies and sapphires are defined below, and examples of some of the clarity grades are seen in figures 7.12 to 7.17.

GIA Clarity Grades
(for Type II colored stones such as ruby & sapphire)

VVS	(Very, very slightly included). **Minor** inclusions: somewhat easy to see under 10X magnification. Usually invisible to the unaided eye.*
VS	(Very slightly included). **Noticeable** inclusions: very easy to see under 10X. Sometimes visible to the unaided eye.
SI_{1-2}	(Slightly included). **Obvious** inclusions: large and/or numerous under 10X. Usually easy to see with the unaided eye: SI_1, visible; SI_2, very apparent.
I_1	(Imperfect). **Moderate effect** on appearance or durability.
I_2	**Severe effect** on appearance or durability.
I_3	**Severe effect on both** appearance and durability.
Dcl	(Declasse). **Stone not transparent.**

* Visibility guidelines are for the trained, experienced observer.

AGL Clarity Grades*
(for all transparent gemstones except diamonds)

FI	**Free of inclusions** with the unaided eye.
LI_{1-2}	**Lightly included** with the unaided eye.
MI_{1-2}	**Moderately included** with the unaided eye.
HI_{1-2}	**Heavily included** with the unaided eye. Inclusions are obvious.
EI_{1-3}	**Excessively included**. Severe effect on beauty, transparency, and/or durability.

* Comments about texture and color zoning are often included with grades on AGL lab certificates. For example, the clarity of the stone in fig. 7.15 would be described as HI_1 with a moderate to strong texture. AGL also assigns split clarity grades (e.g. LI_1 - LI_2) to borderline stones.

Do not assume that everyone who uses AGL or GIA grades applies the same definitions to them. One way of defrauding the public is to assign supposedly high color & clarity grades to poor quality stones. Therefore, ask salespeople to define the grades they use, and always examine the stones yourself both with and without magnification before you buy them.

Fig. 7.12 VVS / LI_1 - LI_2

Fig. 7.13 VS / MI_1

Fig. 7.14 SI_1 / MI_2

Fig. 7.15 SI_2 / HI_1

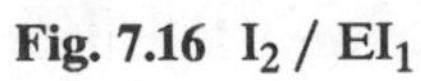

Fig. 7.16 I_2 / EI_1

Fig. 7.17 I_3 / EI_2

Note: The above grades are approximate. Clarity grading cannot be accurately portrayed with enlarged, two-dimensional photographs.

If you're familiar with diamond clarity grades, you might be surprised at how much lower the grading standards are for colored stones. The GIA and AGL systems don't even have a separate grade for colored stones that are flawless under 10-power magnification. That's because although diamonds can be flawless or near flawless, most colored stones are not. Also, unlike diamonds, a flawless colored stone does not command a higher price than the next grade (VVS or FI). In fact, it may be less desireable due to the importance of inclusions in determining if stones are natural or synthetic. When you shop for rubies and sapphires, do not expect the same degree of clarity as you would for a diamond.

You don't need to memorize the above definitions. Just use them as a reference and notice the ways the jewelry trade categorizes gems so that you'll understand price differences. When you judge clarity, also keep in mind that grades can be misleading. In spite of the fact that the GIA diamond grading scale is used internationally, some salespeople may, for example, misgrade an SI_2 diamond as a VS_2.

Clarity grading for colored stones can be even more misleading because there is no one accepted system. There are many grading scales and even the grades used in them can be deceptive. For example, a store might use a system ranging from AAAAA to A with A being equivalent to a GIA I_3 or an AGL E_2.

Instead of asking jewelry salespeople the clarity grade of a colored stone, you might ask them how its clarity compares to other stones of the same gem type in the store. If you're buying a stone for a ring, also ask how durable the stone is. Then look at the stone closely with and without magnification. If you're told that a sapphire with big, eye-visible, brown areas is a store's top quality, you know it only sells lower quality merchandise. If a salesperson tells you that a stone with a long, deep crack across it would be ideal for an every-day ring, you should consider going to another store.

The next two sections will explain how to judge clarity and how to distinguish between flaws that are acceptable and flaws that aren't. If you follow the guidelines below, you will be more likely to get good value on a ruby or a sapphire than if you just rely on the clarity grade a store assigns to a stone.

How to Determine Which Flaws are Acceptable

When deciding if flaws are acceptable, ask these two questions:

- Do the flaws threaten the durability of the stone?
- When a stone is viewed with the unaided eye, do the flaws mar its overall beauty?

If either of your answers is yes, the flaws should probably be avoided. Examples of unacceptable flaws are:

- **Big, deep cracks**. Look at figures 7.18 & 7.19. The stone in these two photos has many serious cracks that make it unsuitable for every-day wear in a ring and extra precautions should be taken when it is cleaned. This stone is a 3.94 carat, purple sapphire that is a cabochon cut (dome-shaped with no facets), and it was bought in California in 1990 for less than $15. If the stone had been faceted and had a lot better clarity, it could have been sold for several times that amount.

  Stones like the one in figures 7.18 & 7.19 might be advertised with headlines like **FOUR-CARAT RUBY: JUST $399. 14-karat gold pendant setting FREE with purchase.** (Purple sapphires are sometimes called rubies since purple is almost red). Even though $399 for this stone would be a rip-off, at $15 it is not a bad buy, especially considering that the cracks don't show when you wear the stone. A stone like this could be worn as an earring or pendant, where it wouldn't be subject to blows and knocks. When you look at ads or shop around, remember that gem prices are meaningless if you don't have specific details about their color, shape, weight, clarity, cutting style, and quality of cut.

- **Big chips** (fig. 7.20). In addition to looking bad, big chips are likely to grow bigger through normal wear.

- **Large, eye-visible, opaque crystals**. Besides blocking the passage of light and reducing brilliance, these crystals may create stress on the stone. This means that if the stone were knocked hard, it could break. Large crystals are sometimes desirable. If they are unusual or attractive they can make the stone a collector's item.

- **Large, white, black or colored areas visible to the naked eye** (fig. 7.21). It may be difficult to tell what these flaws are. They may be crystals, feathers, zoning or some other type of flaw, but they can reduce brilliance and might threaten the stone's durability. If you're looking for a multi-colored stone, these flaws could be acceptable. But if you plan to wear such a stone in a ring or bracelet, first ask your jeweler about it's durability.

  Occasionally, the white areas of stones are just surface abrasions. These can be eliminated or reduced by repolishing.

Fingerprints (partially-healed cracks) were not listed above as unacceptable inclusions. That's because they don't weaken the stone as much as cracks. They can, however, indicate an area that is not as strong as the rest of the stone. Consequently, if there are very large fingerprints extending across the stone, it may not withstand hard wear in a ring. Normally, though, there's no need to reject a stone if fingerprints are present. They are common in corundum, and their unique patterns personalize your stone.

Sometimes laypeople think they should avoid stones with cloudy or misty areas. This is not necessarily true. Silk can look like clouds, but these extremely fine needlelike inclusions are usually microscopic and don't threaten the durability of the stone. They can, however, increase a stone's value by providing evidence that it's from Burma. These tiny needles can also help prove that a stone is not heat-treated or man-made.

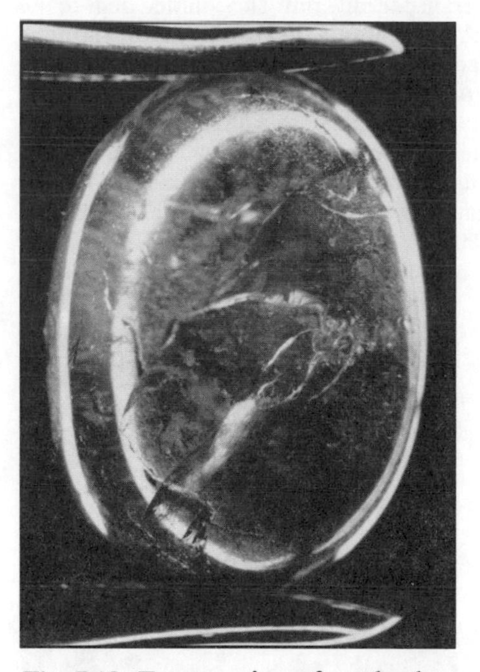

Fig. 7.18 Face-up view of a cabochon with serious cracks

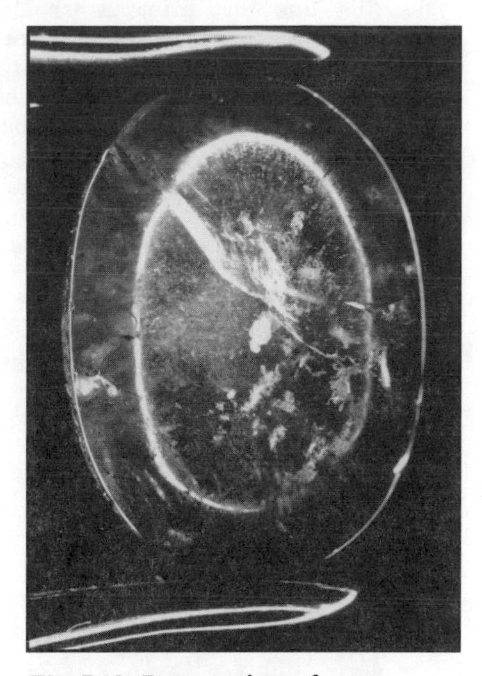

Fig. 7.19 Bottom view of same cabochon

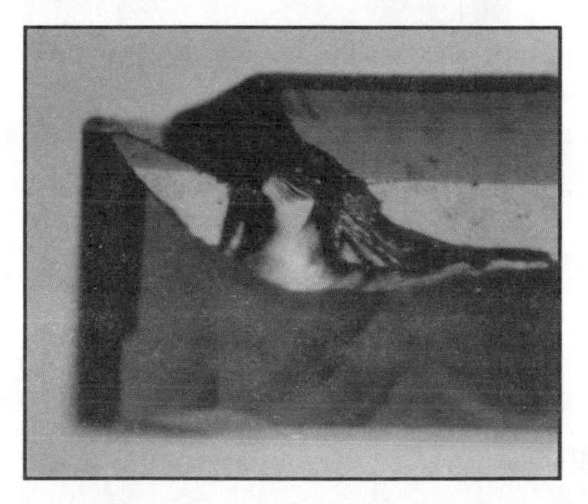

Fig. 7.20 Chipped sapphire baguette

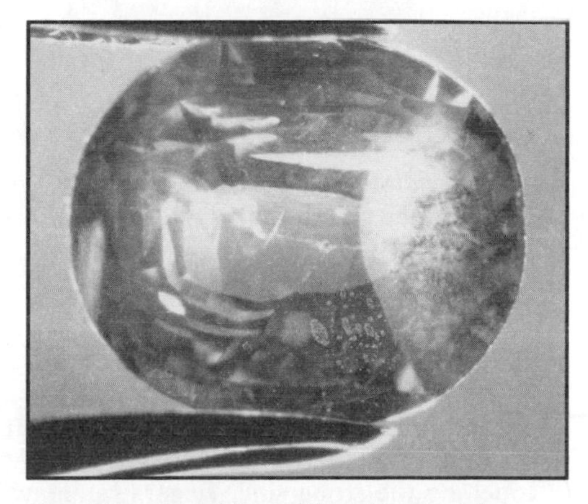

Fig. 7.21 Sapphire with a large white feather area

It's true that diamonds should be highly transparent, but the same degree of transparency is not expected of a ruby or a sapphire. A good illustration of this is the Kashmir sapphire, which besides being prized for its color, is also known for its hazy, velvety look. Even though it's okay for a stone to appear hazy, it's important that light be able to shine through it and reflect back to you. In fact, another way of determining if flaws in a stone are acceptable is to notice if the stone has a high degree of brilliance. If it does, chances are you have a stone that is not only well-cut but is also of good clarity. On the other hand, if the stone is so milky that light doesn't pass through it, it's low-quality, cheap material. An example of this type of nontransparent stone is shown in figures 7.22 and 7.23.

Fig. 7.22 A nontransparent, low-quality sapphire with light shining on it

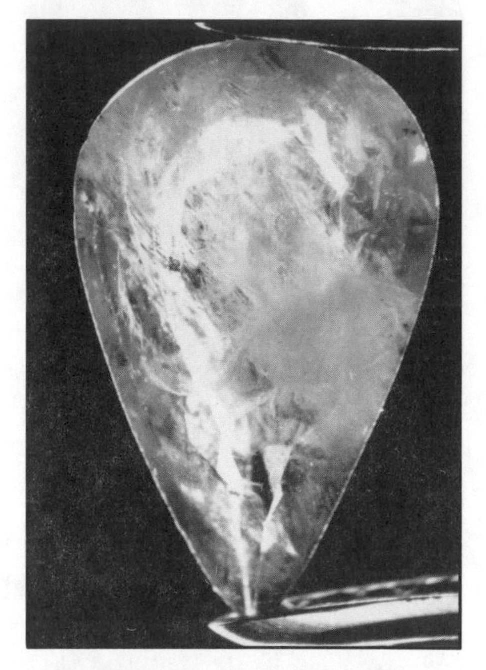

Fig. 7.23 Same stone with light shining through it from the side

Tips on Judging Clarity

- **Clean the stone**. Otherwise you may think dirt and spots are inclusions. Usually rubbing a stone with a lint-free cloth is sufficient. If you're examining jewelry at home, it may have to be cleaned with soap and water. Professional cleaning might also be necessary. Avoid touching the stone since fingers can leave smudges.

- **Look at the stone under magnification.** Even though you should first judge clarity with the unaided eye, you often need magnification to spot threatening cracks. Under magnification, the cracks in figures 7.18 and 7.19 are obvious to a layperson; but with the naked eye, they aren't because the deep color of the stone helps hide them.

Reliable jewelers will be happy to let you use their microscope or loupe (a hand magnifier as seen in figure 7.24). If you're seriously interested in gemstones, you should own a fully-corrected, 10-power, triplet loupe. You can buy these at jewelry or gem supply stores. Plan on paying at least $25 for a good loupe.

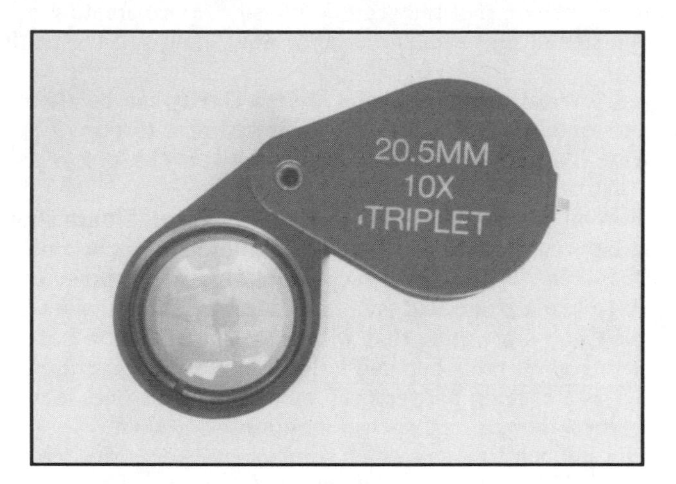

Fig. 7.24 10-power, triplet loupe

- **Look at the stone from several angles**--top, bottom, sides. Even though top and centrally-located flaws are the most undesirable in terms of beauty, flaws seen from the sides or bottom of a stone can affect its price or durability.

- **Look at the stone with light shining through it from the side.** This will help you see flaws inside the stone.

- **Look at the stone with light reflected off the surface.** This will help you identify surface cracks. All the cracks seen on the surface of the stone in figures 7.18 and 7.19 make it an ideal candidate for some red ruby oil. This could improve its apparent color and help hide its cracks (see Chapter 11 for more details on oiling). If you see lots of surface cracks and the stone is set in a jewelry piece, ask the jeweler to clean the piece in his ultrasonic machine for a couple minutes before you buy it. If there's a difference in the clarity or color of the stone after it's cleaned, you may wish to choose another stone.

  Do not, however, ask a jeweler to put an emerald in an ultrasonic cleaner. Since emeralds often have surface cracks, it's a normal practice to oil them. Jewelers themselves will warn you that cleaning emeralds in ultrasonics might change their appearance. Rubies and sapphires are different. Normally they can withstand hard wear and can be cleaned ultrasonically.

- **When you judge clarity, compare stones of the same color.** Not all corundum can be judged by the same standards. Just as it's normal for emeralds to have more flaws than rubies, it's also normal for rubies and padparadschas to have more flaws than other members of the corundum family. Yellow sapphires should have the fewest flaws.

- **Remember that prongs and settings can hide flaws.** Therefore if you're interested in a stone with a high clarity grade, it may be best for you to buy a loose stone and have it set. When jewelers or appraisers tell you that they can only assign approximate grades to stones set in jewelry, this is not a sign of incompetence. They are telling you the truth.

- **Remember that your overall impression of a stone's clarity can be affected by the stones it is compared to.** A stone will look better when viewed next to one of low clarity than next to one of high clarity. To have a more balanced outlook, try to look at a variety of qualities.

- **Tell your jeweler how often and where the stone will be worn.** Stones should fit the purpose for which they will be used. Rubies and sapphires with lots of inclusions but good color can be a good buy for stud earrings whereas they might be a poor choice for an everyday ring.

  If you happen to like a stone that isn't very durable and you want to wear it in a ring, there is a solution. Choose a setting that will protect it such as a bezel setting. This is a band of metal that surrounds the stone and holds it in place. Sometimes a bit of epoxy glue can be added to act as a cushion between the metal and the stone. Setting a very included stone can be dangerous, though, and special care must be taken.

  You may find it difficult to know which settings and stones are strong and appropriate. That's why you should tell your jeweler how they will be worn. A well-informed professional can then help you make a choice that will suit both your needs and your budget.

Chapter 7 Quiz

Select the correct answer.

1. Which is the best type of hand magnifier for examining a gemstone?

   a. 10-power doublet (double lens) loupe
   b. 10-power triplet (triple lens) loupe
   c. 30-power doublet loupe
   d. 30-power triplet loupe

2. Which of the following is the least serious?

   a. Broad color bands
   b. A long crack
   c. A large crystal
   d. A big chip

3. Which of the following are commonly found in rubies and sapphires?

   a. Naturals
   b. Carbon inclusions
   c. Fingerprint inclusions
   d. Laser drill holes

4. You're in a jewelry store and the owner asks to see your ruby ring. He places it under a microscope and tells you the ruby has a small crack in the center so it is a lousy stone. When you look through the microscope, you are able to see a tiny line in the center of the ruby and a small round form near the girdle. This means:

   a. Your ruby is defective
   b. Your ruby will soon crack into pieces
   c. The owner is unprofessional, and he is giving you misleading information.
   d. The owner is a true ruby expert and deserves your patronage.

True or False?

5. Rubies tend to have more flaws than sapphires.

6. Rubies and sapphires don't need to be examined under magnification because what's important is their color and general appearance.

7. Flaws are more obvious in light-colored stones than in darker ones.

8. Flaws can sometimes increase the value of a ruby or a sapphire.

9. The GIA clarity grade "SI_1" has four different definitions.

10. Clarity is the least important factor in determining the price of a ruby or sapphire.

11. When buying a ruby or a sapphire, you should try to find one that is flawless.

12. When examining a ruby or sapphire for clarity, you should look at it from several angles.

13. If a jeweler tells you the clarity grade of a stone and the grade is high, you don't need to look at it under magnification.

Answers

1. B - A triplet loupe is best because it prevents distortion of the image and color. 30-power loupes are rare. 20-power loupes are available, but they are difficult to use and focus because they have such a small depth of field and viewing area.

2. A

3. C

4. C

5. True

6. False - Cracks that could threaten the durability of the stone may be hard to see with the naked eye. Besides being an aid to clarity grading, magnification is also important for detecting imitations, synthetics and treatments.

7. True

8. True - by serving as evidence that a stone is untreated or that it's from a desirable area such as Burma or Kashmir.

9. True - One definition for diamonds and three for colored stones.

10. False - The shape is probably the least important factor. If the clarity is terrible, rubies and sapphires will have a low value no matter how good their color is, how well they are cut, or how big they are.

11. False - A large flawless stone could be harder to resell. Jewelers and dealers tend to be suspicious of flawless rubies and sapphires because that can be a sign the stone is synthetic.

12. True

13. False - Not all jewelers grade stones alike. Some may even change the definitions, and the stone might in actuality have a low clarity. To get a clear and accurate picture of what a gemstone is like, it's best to look at it both with and without magnification.

8

Judging Cut

Cut plays a major role in determining the value of rubies and sapphires because it affects their color and clarity as well as their brilliance. For example, a stone that is cut too shallow can look pale and lifeless, and it can display flaws that would normally not be visible to the naked eye.

The term "cut" is sometimes confusing because it has a variety of meanings. Jewelers use it to refer to:

- The **shape** of a gemstone (e.g. round or oval)
- The **cutting style** (e.g. cabochon or faceted, brilliant or step cut, single or full cut)
- The **proportions** of a stone (e.g. deep or shallow pavilion, off-center culet)
- The **finish** of a stone (e.g. polishing marks or smooth flawless surface, misshapen or symmetrical facets)

The proportions and finish are also called the **make** of the stone. Proportions and how they affect the appearance of rubies and sapphires will be the focus of this chapter. Shape and cutting style were discussed in chapter three. Finish will not be discussed because it normally does not have much of an effect on the price of corundum. If there is a problem with the finish, it can usually be corrected by repolishing the stone. Blemishes such as scratches and abrasions are sometimes considered as part of the finish grade of the stone. In this book, they are discussed in the chapter on clarity.

What is a Poorly Cut Stone?

If you were to ask gem dealers to describe an ideally proportioned ruby or sapphire, you would get a variety of answers. But even though a standardized system for evaluating cut has not yet been adopted, gem experts would agree that a stone is poorly cut if it has the following characteristics:

♦ **An obviously unsymmetrical shape or profile** (fig. 8.1). It's common for rubies and sapphires to look less symmetrical than stones such as diamonds and amethysts. However, when corundum stones are so lopsided that their brilliance is seriously diminished, the lack of symmetry is unacceptable.

A culet that is extremely off center is one of the most serious symmetry problems. It prevents the pavilion from reflecting light evenly. (In cushions, ovals, and marquises, the culet should be centered widthwise and lengthwise. In heart and pear-shaped stones, the culet should only be centered widthwise.)

♦ **An excessive depth** that makes it impractical to set in jewelry (fig 8.3). Some star rubies and sapphires and other cabochons are cut so deep that they almost look like marbles. These stones are impractical for jewelry because they must be set awkwardly high to avoid touching the skin. Also, you end up paying for weight you can't see face up and which does nothing to contribute to the beauty.

Pale, faceted stones are sometimes cut abnormally deep to intensify their colors. Dark stones are cut this way too in order to save weight from the original rough. This unnecessary weight adds to the cost of the stone since prices are calculated by multiplying the weight times the per-carat cost. Consequently, when you compare the prices of stones, you should consider their overall depth.

♦ **A very lumpy, bulging pavilion** (fig. 8.2) It decreases brilliance and helps create dark or window-like areas in the stone. This is because the pavilion is not slanted at an angle that will maximize light reflection. The bulging pavilion is common in corundum, and it is another example of how you can end up paying for extra weight that only reduces the beauty of the stone.

♦ **A big window area**, which allows you to see right through the stone. Windows can occur in any transparent, faceted stone no matter how light or dark it is and no matter how deep or shallow its pavilion is. Improper crown and pavilion angles are the main cause of windows. In all colored stones, these windows are undesirable because they mean reduced brilliance and color.

To look for windows, hold the stone about an inch or two (2 to 5 cm) above a contrasting background such as your hand or a piece of white paper. Then try to look straight through the top of the stone without tilting it, and check if you can see the background or a light window-like area in the center of it. If the stone is light colored, you might try holding it above a printed page to see if the print shows through.

Fig. 8.1 An off-center culet of a marquise-shaped ruby

Fig. 8.2 A lumpy pavilion

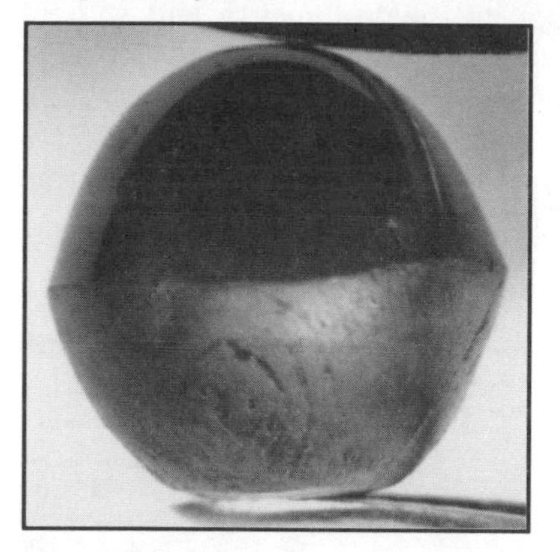

Fig. 8.3 A stone that is too deep

Fig. 8.4 "Window" in a deep stone

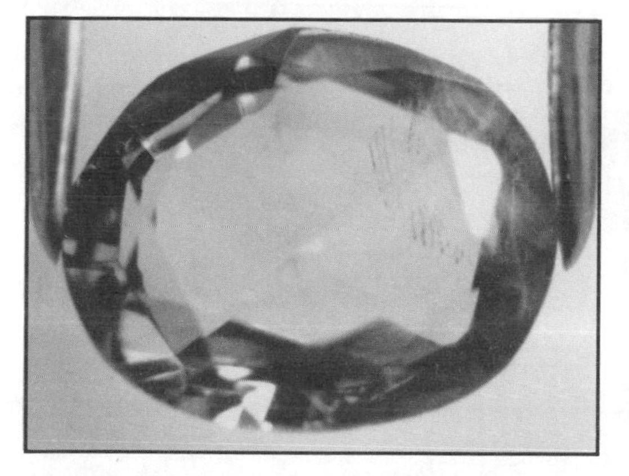

Fig. 8.5 Large "window" in a shallow stone

Fig. 8.6 "Window" in a stone with an off-center culet

Sometimes, when a gem with a large window is shown to you in a stone paper or on a pad, it may appear well-cut. This is because the stone is right next to the background, and there is not enough light underneath it to expose the window. That's why it's important to examine stones **above** a contrasting background.

It's common for rubies and sapphires to have a slight window. However, windows like the ones in figures 8.4 to 8.6 would be unacceptable to someone looking for a very brilliant stone with intense color.

Judging the Profile

When you buy a ruby or sapphire, be sure to look at its profile with and without magnification. The side view can indicate:

- If the stone is suitable for mounting in jewelry.
- If the stone will look big or small for its weight.
- If the cutter's main goal was to bring out the stone's brilliance.

Figure 8.7 reviews some fundamental gem terminology and serves as an example of a well-proportioned mixed-cut ruby or sapphire.

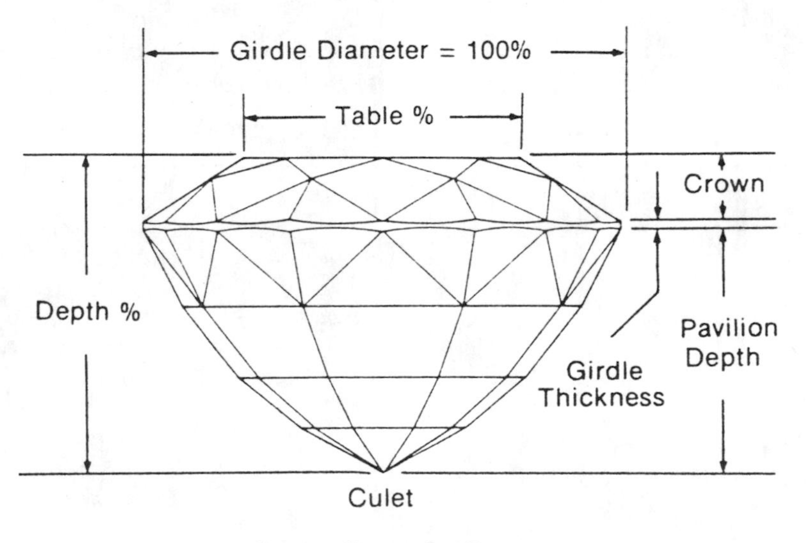

Fig. 8.7 Profile diagram of a mixed-cut colored stone. Copyright by American Gemological Laboratories, Inc., 1978.

The **depth percentage** of the stone in figure 8.7 can be calculated as follows:

depth
width (girdle diameter)

In this case the depth percentage is 65%, which is a good depth/width ratio for a colored stone. There are differences of opinion as to what is the best depth percentage for a ruby or a sapphire. Some say 60 to 65%. Others say 70 to 80%. Combining these two ranges, we can conclude that a stone's depth should be about 60 to 80% of its width.

If a stone is much deeper when you look at it widthwise, then it may not be suitable for mounting in jewelry and it will look small for its weight in the face-up view. The main reason for cutting extremely deep stones is to save as much weight from the original rough as possible. Stones may also be cut deep to darken their color, especially if they are pale or color-zoned.

If a stone is extremely shallow ("flat") when you look at it widthwise, it might be fragile and therefore unsuitable as a stone for an everyday ring (it could, however, be ideal for a pendant or drop earrings). Very shallow stones look big for their weight in the face-up view, but unfortunately they often have big windows and lack brilliance, which brings down their value. The main reason for cutting extremely shallow stones is to save as much weight from the original rough as possible. Stones may also be cut shallow to lighten their color.

When judging the profile of a ruby or sapphire you should also **pay attention to the crown height and the pavilion depth**. Notice the relationship of the crown height to the pavilion depth in the diagram of figure 8.7 (about 3.5 to 1). Then compare the profile of figures 8.8 - 8.10 to the diagram. Without even measuring these stones, you can make visual judgments about their pavilion and crown heights.

If the crown is too low, the stone will lack sparkle (figs. 8.8 & 8.9) When light falls on a flat crown, there tends to be a large sheet-like reflection off the table facet instead of twinkles of light from the other crown facets.

If the crown or pavilion is too flat or too deep, the stone may lack brilliance, have a window, or look blackish. In order for the stone to effectively reflect light, the crown and especially the pavilion must be angled properly. But they can't have the proper angles if they don't have the proper depth.

Judging Brilliance

When the term **"brilliance"** is applied to colored stones, it means their ability to reflect light and color back to the eye. If a stone is brilliant, it has **"life."** If the stone has very low brilliance or none at all, it is **"dead"**. Brilliance should not be confused with **glare**, which is a

Fig. 8.8 A very low crown and an off-center culet of an oval ruby

Fig. 8.9 Another low crown

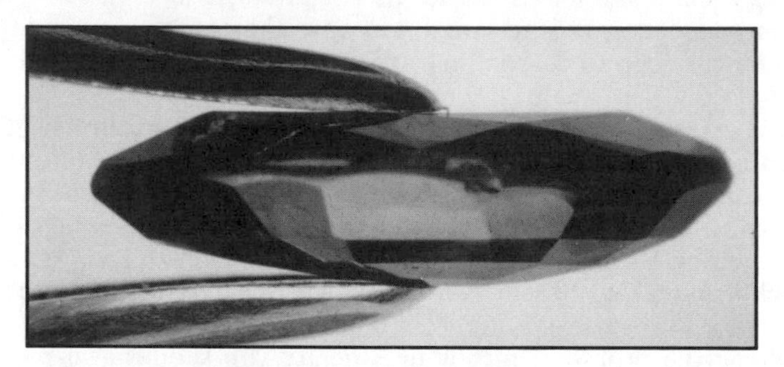

Fig. 8.10 An example of a low pavilion

Fig. 8.11 Profile of a sapphire with an acceptable crown height & pavilion depth

Fig. 8.12 Another acceptable profile

pale or white reflection of light off the surface of the stone. To better understand the concept of brilliance, look at the stones in figures 8.13 to 8.20 and notice the varying degrees of brilliance.

Another way you can learn to recognize brilliance is to look at some blue topaz and sapphire jewelry in a few stores. Compare the amount of color reflected from the facets of the stones. In some stores, you may have a hard time finding many blue reflections in the sapphires. These stones may instead appear to have one solid color--very dark navy blue. There are lots of these dark, non-brilliant sapphires in jewelry showcases because the stones are inexpensive. Stores are trying to meet their customers' demands for low-priced jewelry.

The blue topaz stones you see will often look more brilliant than the sapphires. Topaz rough costs a lot less than sapphire, therefore cutters don't need to be so concerned about making the stone as heavy as possible. They can cut the topaz to bring out its beauty. Another reason for the difference in brilliance is that lighter-colored stones tend to be more brilliant than darker ones.

After you have compared the topaz and sapphire jewelry, try to pick out which sapphires look the most brilliant. Look for bright blue reflections in the stones. If you were trying to select a brilliant ruby, you would look for flashes of bright red. Keep in mind that it's normal for these stones to have some dark and non-brilliant areas. It's rare for more than 80% of a ruby or sapphire to look brilliant.

When comparing brilliance, also follow these guidelines:

- Examine the stones against the same type of background and the same distance away from the background.
- Look straight down on the top of the stones.
- Turn the stones as if they were a screw and look at them lengthwise and widthwise. The bright colored areas will probably move around.
- Look for windows in the stones and compare the sizes of these windows. It will be easier to spot windows if you look at the stones above both white and black backgrounds.
- Compare the dark or black areas in the stones. (In the trade these are called areas of **extinction**).
- Compare the stones under the same kinds of light. The degree of brilliance varies according to the type, position, and intensity of the light.
- Look at the stones away from direct light and check how much brilliance and color still remains.

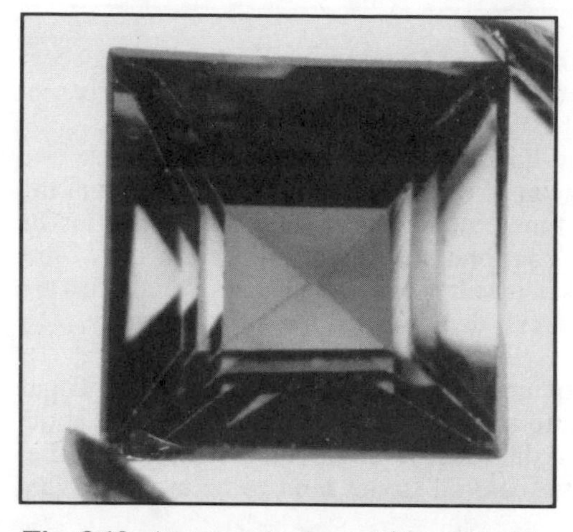

Fig. 8.13 A step-cut, square, blue sapphire viewed against a white background. Note the light-colored areas including the light square area in the center of the stone.

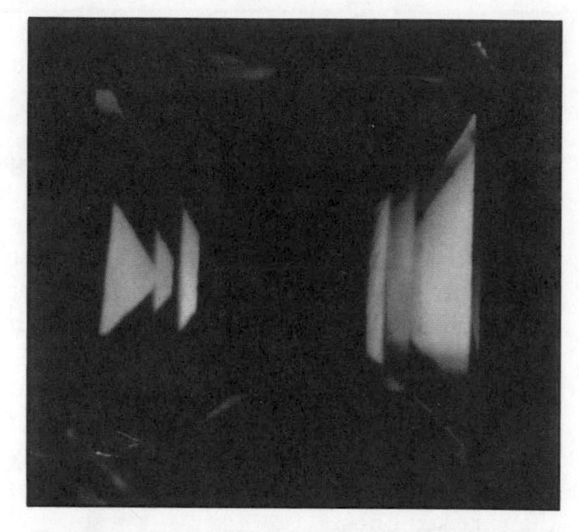

Fig. 8.14 Same stone viewed against a black background. The central square is now black, which indicates it is a nonbrilliant "window" area. The remaining light areas are brilliant areas. This stone does not have a high degree of brilliance, but it has more than a lot of stones of its type.

Fig. 8.15 Another step-cut blue sapphire. Notice how the light-colored areas are scattered around the stone. When this stone is turned, it shows more sparkle than the stone above.

Fig. 8.16 Note how the light (brilliant) areas extend into the center of the stone even when viewed against black. For a small, step-cut sapphire, this stone has a lot of "life."

Fig. 8.17 A brilliant-cut yellow sapphire (0.44 ct princess cut). Light-colored and brilliant-cut stones have a greater potential for brilliance than step-cuts or darker stones.

Fig. 8.18 Even against a black background this stone displays a lot of color due to its numerous brilliant areas. This stone has "life."

Fig. 8.19 This blue sapphire with its numerous facets has a good play of light even though both the crown and pavilion are step cut.

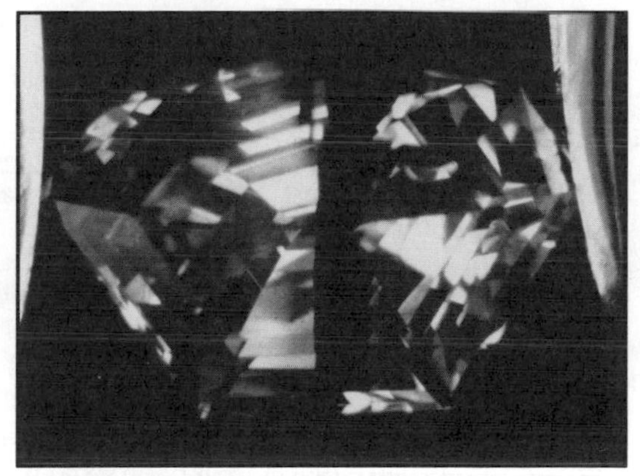

Fig. 8.20 Note how the brilliant areas extend close to the center, indicating the absence of a large, undesirable "window" (a common problem in rubies and sapphires). The proportion of brilliant areas to black (extinction) areas in this sapphire is acceptable, even though it isn't a stone of exceptional brilliance.

How Cut Affects Price

Theoretically, major cutting defects should reduce corundum prices substantially. In actual practice, this is not always true.

Sometimes the cut may have no direct effect on the per-carat price. For example a 3/4-carat sapphire pendant may be mass-produced and sold at the same price in chain stores. Some of the sapphires may have a good deal of brilliance. Others of similar color and clarity may have almost no brilliance. In spite of their identical price, the appraised value of the more brilliant and intensely-colored sapphires should be greater.

In some stores, you can select a ruby or sapphire from an assortment of stones in a packet or little bowl. The per-carat price for all the stones may be the same even though they might vary considerably in quality. People that know how to judge cut, color, and clarity will get the best buys in cases like these because they will be able to pick out the most valuable stones and avoid paying for unnecessary weight.

Sometimes the cut has a mathematically calculated effect on the price of corundum. For example, the prices of fine-quality princess cuts that are calibrated to specific millimeter sizes may be determined by the weight lost when cutting the stones.

If you were to have a large stone recut to bring out its brilliance, you could calculate its new per-carat price by dividing the new weight into the combined cost of the recutting and the stone. The equation for this is:

$$\text{New per-carat cost} = \frac{\text{stone cost} + \text{cost of recutting}}{\text{new carat weight}}$$

In spite of the recutting costs and lost weight, the new appraised value of a recut stone might be considerably higher than the actual new per-carat cost. Consequently, it is not uncommon for fine-quality corundum material to be recut to improve the proportions.

Sometimes the cut affects corundum prices in a subjective manner. Some dealers place a greater importance on cut than others and they may discount a very poorly cut stone as much as 50% in order to sell it. Another dealer might discount the same stone 25%. There is no established trade formula for determining percentage-wise how cut affects the value of corundum. There is, however, agreement that a well-proportioned, brilliant stone is more valuable than one that is poorly cut.

The lack of formal guidelines on how cut affects stone pricing may be dismaying, but it can work to your advantage. Due to the flexibility of prices, you may be able to find high quality stones at a much lower cost than normal.

Finding an attractive stone, however, is more important than finding one at a bargain price. Rubies and sapphires with lots of life and color will always be valuable, but they're not always readily available. This is unfortunate because man has control over a stone's cut. There could be a better selection of well-cut corundum stones, but the public has to ask for this.

As you browse in jewelry stores, make a special effort to notice the degrees of brilliance and color in the rubies and sapphires. This helps develop the eye. Once you can recognize a fine-cut, brilliant stone, you'll be well on your way to spotting value.

Chapter 8 Quiz

True or False?

1. The way a gemstone is cut can affect its color, clarity, and brilliance.

2. Even though the actual price of a ruby or sapphire may vary, gem dealers normally agree on what percentage a poorly cut stone should be discounted.

3. A good sapphire should look equally brilliant under different types of light.

4. Since rubies and sapphires are so expensive, cutters usually aim for maximum brilliance and precision proportioning when cutting these stones.

5. Rubies and sapphires with low pavilions tend to have "windows."

6. Light-colored stones tend to be more brilliant than those with darker colors.

7. Rubies and sapphires with deep, bulging pavilions usually look large for their weight.

8. Sometimes poorly cut rubies and sapphires in a parcel of stones are priced the same as the better cut stones in it.

9. The best way to judge brilliance is to look at the profile view of a stone to see if it is proportioned properly.

10. Stones with few facets and very low crowns generally have less sparkle than those with more facets and higher crowns.

Answers

1. True

2. False - Both the price and percentage of discount can vary from one dealer to another.

3. False - The amount of brilliance even in high-quality stones varies depending on the lighting under which the stone is viewed.

4. False - Unlike diamond cutters, who are also dealing with expensive gem material, ruby and sapphire cutters tend to aim more for maximum weight than brilliance. As a consequence, there are often large dark areas and "windows" in rubies and sapphires. This in turn means reduced color and brilliance.

5. True

6. True

7. False - They usually look small because the weight is concentrated below the stone where you don't see it face up.

8. True

9. False - The brilliance of rubies and sapphires is also affected by their color, clarity, and cutting style. Stones with a great deal of silk, for example, will be less brilliant than if they were more transparent. To judge how all of these factors affect the brilliance of the stone, you have to look at it in the face-up position.

10. True

9

Imitation or Real?

Imagine that you've just found a ring with a red stone on the beach. You wonder if you should have it appraised. You're afraid, however, that the ring might be worth less than the cost of the appraisal.

In this and other situations, such as at flea markets or garage sales, it would be helpful to be able to make an educated guess about the identity of a gem. The guidelines in the following section can help you determine whether a stone is a ruby or sapphire.

Tests a Layperson Can Do

Color Test

Compare the color of the stone to that of other rubies and sapphires you've seen. Is there a dramatic difference? If so, it could be a synthetic or glass.

Color should never be the sole basis for identifying a gemstone. However, it is an important consideration. When browsing in jewelry stores, notice the various shades of color of the rubies and sapphires. This will increase your color awareness, and in turn, it will be easier for you to identify these stones.

Size Test

Note the size of the stone. Rubies over three carats are rare. Consequently, if you see a very large red stone, say 3/8 inch (10mm) in width, at a flea market for example, chances are it is not a natural ruby. Blue sapphires weighing more than 5 carats are not common, so be suspicious too of blue stones in very large sizes.

Shape and Cutting Style Test

Rubies and sapphires of a carat or more can be found in a wide variety of shapes, but the most common are oval and cushion (see Chapter 3). Whatever the shape, they're almost always a mixed cut. Round stones, however, are usually brilliant cut.

Occasionally large stones are emerald cut, but be wary when you see one. If you're looking at a large, round, single-cut stone (with just 9 facets on the top), you can almost be assured that it is glass or some other sort of imitation (fig. 9.1). Shape and cutting style are never proof of a stone's identity, but they can serve as good clues that a stone is an imitation.

Closed Back Test

If the stone is set in jewelry, look at the back of the setting. Is the pavilion (bottom) of the stone blocked from view or enclosed in metal (figs. 9.2 & 9.3)? Normally, the bottom of a faceted ruby or a sapphire is at least partially visible. Therefore, if you can see only the crown or top of the stone, you should be suspicious.

An open-back setting does not indicate that a stone is a genuine ruby or sapphire. Glass imitations are often set with the pavilion showing. But a completely closed back is often a sign that something is being hidden. Maybe the stone has a foil back or metallic coating to add color and brilliance. Maybe the stone is made of two separate pieces of material that have been glued together. No matter what might be hidden, to avoid being duped, it is best to buy stones in a setting with part of the pavilion showing if you're not dealing with someone you know and trust.

Recently, some jewelry manufacturers have used solid backs under channel-set stones to increase the rigidity of the channel mountings. This allows the stones to be set more securely. In most cases, however, it is still customary to set rubies and sapphires with part of the pavilion showing.

The Price Test

Is the stone being sold at an unbelievably low price? If it is, it might be an imitation, synthetic, or stolen or defective merchandise. Even gem dealers rely on the price test to help them avoid being fooled by synthetics (man-made duplications of gems). They realize that a supplier can't stay in business if he sells stones below his cost.

The Perfect Clarity Test

Does the stone look flawless when you look at it with a 10-power magnifier such as a loupe? If it's a large red stone, it's probably an imitation or a synthetic because a large natural ruby would normally have at least a small inclusion. It's a lot easier to find sapphires of high clarity, but you should be leery of large flawless sapphires when they're not accompanied by laboratory identification reports.

Fig. 9.1 Single-cut glass stone set in cheap metal. The pocked surface and bubbles are typical of glass.

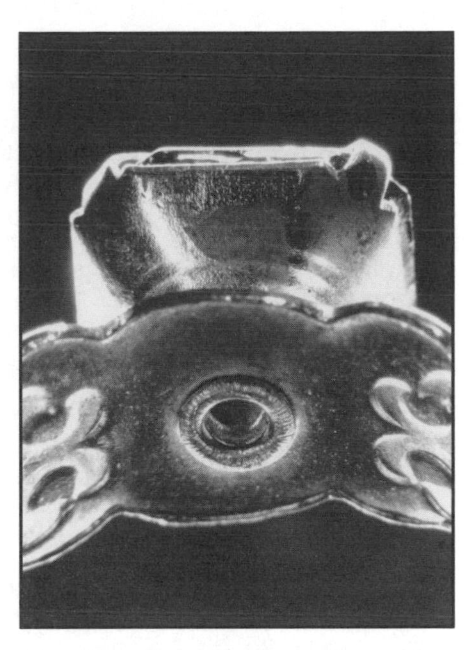

Fig. 9.2 Closed-back setting with a false opening, blocked by metal

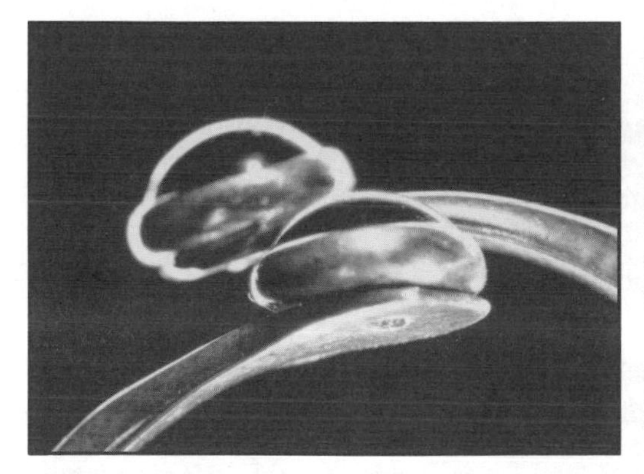

Fig. 9.3 Closed-back settings of small glass cabochon stones

Fig. 9.4 Open-back settings

The Glass Test

One of the most common imitations of corundum is glass. It will be easier for you to recognize it if you learn some of its characteristics. Look at figures 9.5 to 9.8. Notice the following characteristics which are indicative of glass:

- Gas bubbles. In glass they are round, oval, elongated, or shaped like donuts.
- Rounded facet edges. Real gems normally have sharper, more defined facet edges.
- Concave facets and surfaces.
- Swirly lines or formations.
- Uneven or pitted surfaces. In some cases, the surface may resemble an orange peel.
- Very simple faceting styles such as step cuts with 5 crown facets on large stones. It would be surprising to see a one-carat ruby or sapphire cut this way. However, inexpensive gems or very small rubies or sapphires may be cut this simply.
- Heavily abraded areas. Glass is much softer than corundum so it is more easily scratched and scraped. Nevertheless, abrasions are not a very accurate indication of glass because they can appear on any stone, even diamond.

The best way to learn to recognize glass, is to start looking at it closely. Look at some inexpensive drinking glasses with a loupe. There will probably be some bubbles and often they will be visible with the naked eye. Large bubbles are one of the most reliable indications of glass. Look at cheap costume jewelry with a loupe whenever you get a chance and try to find the characteristics above. The more you examine glass, the better you will become at identifying it.

The Crossed Polaroid Test

This test is done by placing a stone between two Polaroid lenses or filters which have been rotated to prevent light from passing through them (crossed-Polaroid position) (fig 9.9). Because of the crystal structure of corundum, light that enters it is split into two beams which travel at two slightly different speeds at right angles to each other. (The technical term for this optical property is **double refraction**). **Doubly-refractive stones** such as ruby and sapphire will turn light and then dark as they are rotated between crossed Polaroid plates (figs. 9.10 to 9.13).

Glass and the gemstones garnet and spinel, which are often mistaken for corundum, are **singly refractive**. This means light passes through them at one speed and in all directions. When singly refractive stones are rotated between crossed Polaroids, they normally remain dark. Sometimes, however, they show what is called anomalous or false double refraction, meaning that they may appear light and then dark as they are rotated between crossed Polaroids. Usually this light and dark blinking is not as distinct as it is with true doubly refractive gems such as rubies and sapphires.

Fig. 9.5 Bubbles in glass with a swirled background formation

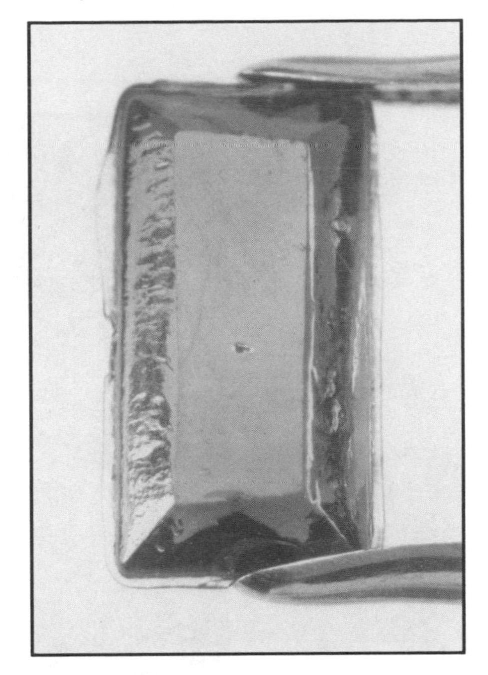

Fig. 9.6 A typical glass stone--rounded facet edges, orange-peel-like surface, a bubble in the center, simple cutting style with just 5 crown facets

Fig. 9.7 Notice the bubbles, the rounded girdle, and the lack of sharp facet edges on this glass stone.

Fig. 9.8 The concave table facet and the rounded facet edges are sure signs that this is a glass stone. Note, too, how the crown has only five facets.

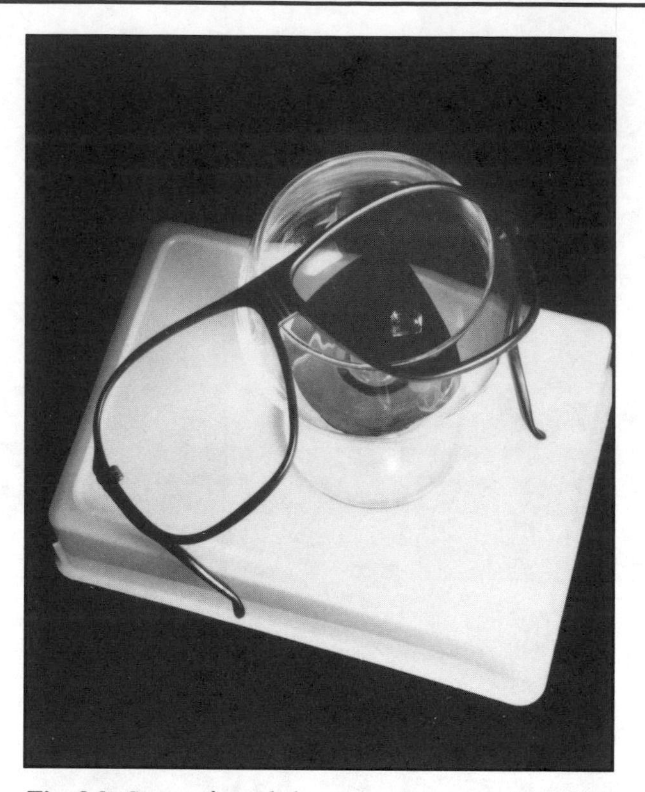

Fig. 9.9 Stone viewed through a home-made polariscope--a wine glass and the lenses of a cheap pair of polaroid sunglasses

Gemologists do the crossed Polaroid test with an instrument called a **polariscope** (The cost is about $180). Basically, the polariscope consists of two Polaroid filters mounted in metal above a light. If you have an old pair of Polaroid sunglasses that you don't mind scratching and a brandy or wine glass with a wide rim, you can do the crossed Polaroid test at home. Just pop one of the lenses out of the sunglasses (this is usually easy to do with cheap plastic or acetate glasses). Old polarizing filters for cameras can also be used. Then follow this procedure.

1. Place the popped-out lens in the bottom of the glass.
2. Put the stone on the lens which is in the bottom of the glass. It's best to position the stone on its pavilion (bottom). If it's mounted in jewelry, then hold it at an angle over the lens. Natural rubies and sapphires lying in a face-down position as in figures 9.12 and 9.13 usually do not show the light to dark blinking.
3. Place the sunglasses with the remaining lens over the rim of the wine glass as in figure 9.9.
4. Get a desk-lamp or spot-light and aim it down at the bottom of the stem so that light will reflect up through the bottom of the glass and Polaroid lens.

Fig. 9.10 Note how dark this stone is when being viewed through crossed polaroid filters.

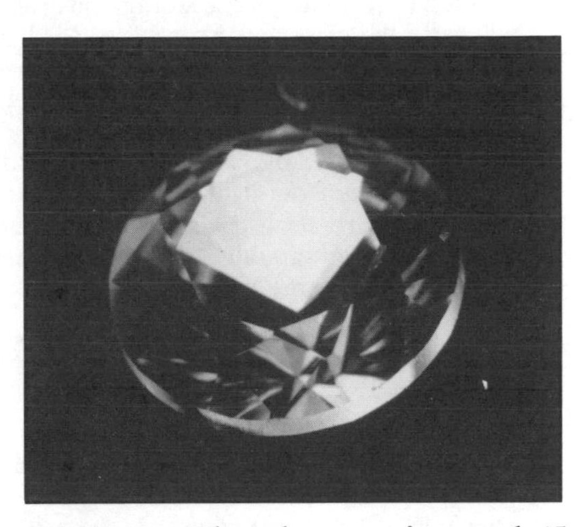

Fig. 9.11 When the stone is turned 45 degrees, it becomes light. This is typical behavior of a ruby or a sapphire but not of stones like glass and garnet.

Fig. 9.12 The same stone, dark and with its table face-down. Normally, natural corundum doesn't change from dark to distinctly light through crossed polaroids in the face-down position.

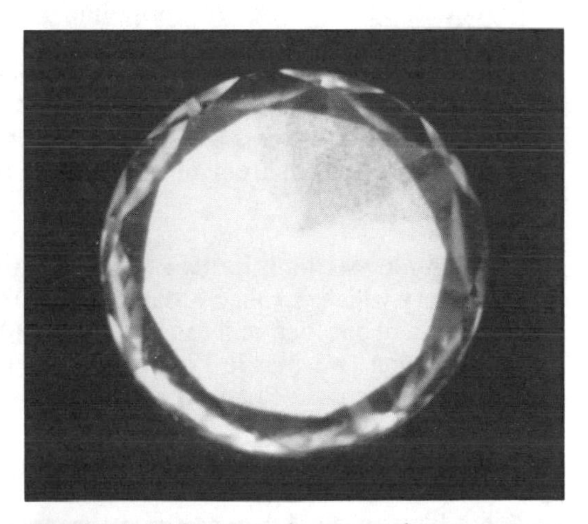

Fig. 9.13 This stone, however, becomes very light when turned 45 degrees. It happens to be a synthetic pink sapphire. Synthetic corundum often reacts this way in the face-down position.

5. If you have a white plastic storage box (like Tupperware), place the glass on top of it. This will raise the glass making it easier for light to come through the bottom.
6. Rotate and position the sunglasses so that when you look through the two lenses, which are parallel to one another, they look black with no light shining through. This is the crossed Polaroid position.
7. Rotate and move the stone using the end of a spoon or other long object. If the stone doesn't change from light to dark as you look at it between the lenses, place it in another position and rotate it again. **If a transparent stone stays dark in all positions, the stone is not a ruby or sapphire.** If it blinks from light to dark, it may or may not be corundum. If the stone stays bright in all positions, it could be badly flawed corundum or another gem.

For this test to work effectively:

1. The light must come through the bottom of the glass instead of the side.
2. The two Polaroid lenses must be in the darkest position.
3. The stone must be rotated 360 degrees in at least two different positions before you can conclude it is not corundum. To be safe, you should check as many positions as possible.

Two-Color Test

When you look at a ruby at different angles, you may see both purplish red and orangy red colors. This is because corundum is doubly refractive and splits light into two rays perpendicular to one another. When gems show a combination of two colors, they are said to be **pleochroic** (multicolored) or more specifically **dichroic** (bicolored). These two dichroic colors can be seen separately through two Polaroid filters placed side by side. However, the light in one filter must pass at right angles to the light in the other filter. In other words, if you slide the two filters in front of each other, they would be in the crossed Polaroid or black position.

Gemologists look for two colors in stones through a cylindrical instrument the size of a AA battery which is called a **dichroscope**. One type is made with Polaroid material (about $40 to $50) and another with calcite (about $100 to $115). You can make a type of dichroscope by scotchtaping two plastic Polaroid lenses together. Place the two lenses in front of one another in the crossed-Polaroid position and then slide them so they are side by side and tape them.

You could also go to a photography store and buy a square sheet of Polaroid acetate (called a filter or a gel depending on the store) for about $15 to $20. Cut a narrow strip from it about three inches (8 cm) long. Then cut the strip in half and tape the two pieces perpendicular to one another with transparent tape as in figure 9.14. The rest of the Polaroid sheet can either be used for the crossed-Polaroid test or for covering lights to cut glare while taking photographs.

Fig. 9.14 Two-color test--Stone viewed through perpendicular polaroid strips scotch-taped to each other

To do the two-color test follow the procedure below:

1. Hold the stone between your forefinger and thumb or in a tweezers.
2. Look through the stone in front of a window or in front of a light covered with some type of white shade (try to avoid fluorescent lighting because it can have a partial polarizing effect).
3. When you can see strong light coming through the stone, place the perpendicular Polaroid strips in front of and against the stone as in figure 9.14. Note: The light you need for the two-color test should be coming through the stone, rather than reflected from a facet.
4. Gradually move and rotate the strips around the stone until you can see one color through the vertical strip and a different color through the horizontal strip. **If the stone is a ruby, it will have both a purplish-red and orangy-red color. If it's a blue sapphire, it will have a greenish-blue and violetish-blue color**. However, the lighter a stone's color is, the less obvious the two dichroic colors are. In certain positions, the two colors will show side by side. In others, it's easier to see the two colors by shifting the Polaroid strips back and forth and viewing one color changing into the other color.

   Singly refractive stones such as glass, garnet, and spinel will show only one color. A doubly refractive stone such as red tourmaline will have different pleochroic colors--red and light red.
5. Make certain that you view the stone through the Polaroid strips from a variety of positions and angles--through the sides, top, bottom, back, and front of the stone. In some stone positions, the two colors will be only barely or not at all visible.

In very small rubies and sapphires, it's easier to see the dichroic colors side by side with a true dichroscope because it has a kind of magnifying effect as you look through it. But even the true dichroscope has its limitations. When stones are mounted in jewelry, it may be impossible for adequate light to show through the stones or it may be hard to position the stones so the two colors show. Consequently, do not automatically assume that a stone is not a ruby or a sapphire if you can't find its dichroic colors. However, if you find the correct dichroic colors, you can assume that you have either a natural or synthetic ruby or sapphire. One of the best ways to learn to recognize these colors is to buy an inexpensive synthetic ruby and sapphire and practice this two- color test on them.

Other Ways of Identifying Rubies & Sapphires

Many of the other identification tests require special training and/or equipment. Nevertheless, you may be curious about how gemologists identify corundum. Therefore, other methods are listed below:

- The **refractive index** (the degree to which light is bent as it passes through the stone) is determined. This is measured with an instrument called a refractometer. Rubies and sapphires have a refractive index of 1.762 - 1.770, which means they bend light 1.77 times more than air does. This also means that light travels 1.77 times more slowly through corundum than it does through air.

  There is no other doubly-refractive gem material that has the same refractive index as corundum. Consequently, the refractometer is a very useful tool for identifying ruby and sapphire, especially yellow sapphire, whose pleochroic colors are similar to those of other yellow gems.

- The interior and exterior of the stone are examined under magnification. Certain features such as those listed in the chapter on clarity are characteristic of ruby and sapphire.

- The stone is viewed in an intense light through a colored filter called a chelsea filter. This test is most helpful for identifying rubies. Synthetic rubies and fluorescent natural rubies like those from Burma have a strong red glow when viewed through a chelsea filter. Other red gems don't look the same.

- A light is directed through the stone with an instrument called a spectroscope to measure how it absorbs light. Corundum has characteristic readings, which are listed in the appendix.

- The stone is placed under short-wave and long-wave ultraviolet light and compared to other gem species of the same color. Descriptions of corundum fluorescence are given in the appendix.

Distinguishing ruby & sapphire from other gem species and glass is not hard if a combination of the preceding tests are done correctly. What can be hard, however, is to separate natural from synthetic corundum. This will be the focus of the next chapter.

10

Synthetic or Natural?

When you think of the term "synthetic," what's the first thing you associate it with? Perhaps rubber? Perhaps textiles like rayon, polyester, or acetate? These are some of the best-known types of synthetic products; and in each case, the synthetic has a different chemical composition and name from the natural material it replaces, whether it be rubber, silk, or cotton.

In the jewelry trade, the word **synthetic** is used differently. It describes a gemstone made in a lab which has the same chemical, optical, and physical properties as its natural counterpart. **A natural gemstone** comes from the ground and is a product of nature, not of man. **Imitations**, on the other hand, do not have the same chemical composition as the stones they resemble, and they may be made by nature or by man. Glass and synthetic spinel, for example, are man-made imitations of corundum. Garnets used to mimic rubies would be natural imitations.

Since consumers tend to interpret the word "synthetic" differently than jewelers, people who sell synthetic rubies and sapphires usually prefer to describe them with terms such as **created, lab-grown, cultured**, or **man-made**. This book, however, uses the term "synthetic" as a way of distinguishing these stones from lab-grown imitations such as cubic zirconia, which imitates diamond.

Synthetic gems are not just a recent phenomenon. Synthetic corundum has been sold commercially since the early 1900's; so if your grandmother has some ruby jewelry, the stones could have been made in a laboratory. Today, synthetic corundum is even more common, especially in birthstone jewelry and class rings. It's also found in designer jewelry, set with diamonds in gold or platinum.

Perhaps you're wondering how anyone could tell the difference between a synthetic and a natural ruby. Even though they each have the same chemical composition, their flaws can be very different. They may also be cut differently. And the coloring agents used in them may cause them to absorb light differently than natural stones.

The following section offers a few tips on identifying synthetic corundum. While reading it, keep in mind that certain types of synthetics require a great deal of expertise and technical equipment for their detection. That's why it's important to deal with reliable and highly trained jewelers when making a major ruby or sapphire purchase. The difference in value between a synthetic and a natural stone can be thousands of dollars.

Tests a Layperson Can Do

Some of the warning signs of imitation stones are valid for synthetics as well. An exceptionally large size, an unnatural color, a closed-back setting, a flawless clarity, and a price that's too good to be true can indicate that a stone might be a synthetic. However, tests that measure refractive index, hardness, and density will not separate the two because these characteristics are the same for both natural and synthetic corundum.

Other ways of detecting synthetic rubies and sapphires are as follows:

Shape and Cutting Style Test

As mentioned in the previous chapter, rubies and sapphires of a carat or more are most commonly fashioned into oval and cushion mixed cuts. Large natural stones that are round or emerald cut are available, but they are harder to find. Synthetic corundum, however, is commonly cut into large round-brilliant and emerald-cut stones. These two cuts, therefore, can serve as a warning signal.

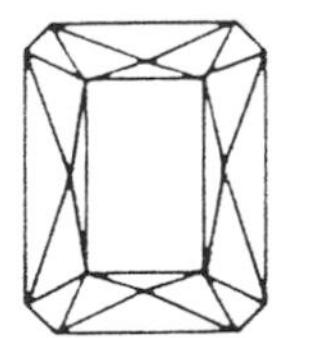

Scissor cut

Another style of cutting that is used on synthetics is the scissor cut, which is used for rectangular-shaped stones. The facets of this cut form an 'X' pattern on all four sides of the crown. If you see the scissor cut on a stone identified as a ruby or a sapphire, you should be very suspicious. It is not likely to be natural corundum.

The Curved Line or Band Test

Synthetic ruby and color-change (alexandrite-like) sapphire are the easiest types of synthetic corundum for a layperson to detect with this test. Curved lines that look like grooves in a phonograph record and that extend across facets can often be seen in these synthetics with a good ten-power loupe (fig 10.1). When present, these lines are proof that the stone is a synthetic.

If you're seriously interested in learning to spot a synthetic, buy a cheap synthetic ruby (preferably an oval or emerald cut at least 6 x 8mm with a strong red color but not a lot of black areas). Inexpensive-type synthetic rubies sometimes sell for less than $5 each at jewelry supply stores or gem & mineral shows.

Using a good, 10-power, **triplet-lens** loupe, try to find curved, groove-like lines in the synthetic ruby. These lines are often visible directly through the table facet; but to find them, you will probably have to look at the stone from several directions with light going through and bouncing off of different facets. A good lamp or daylight through a window are both acceptable light sources. Try aiming the light through the side of the stone. Afterwards, aim it through the bottom. If you can't find the curved lines, ask a jewelry professional to help you. Sometimes, if you first see the lines through a microscope, it's easier to find them with a loupe. It's possible too that the lines are not very distinct in your stone. Once you know how to find these lines in an actual stone, it will be a lot easier for you to detect synthetics. Curved, groove-like lines are present in most synthetic rubies and synthetic alexandrite-like sapphires. They should not be confused with polishing marks, which do not extend across the facet edges.

Fig. 10.1 Curved growth lines in synthetic ruby

Finding curved bands or lines in synthetic blue sapphires or light-colored stones normally is not easy for a layperson. Even professionals may have a hard time finding the bands in these stones. Colored filters and immersion of the stones in special liquids may be needed to make the bands visible. If you can't find curved bands or lines in a stone that you think might be a synthetic, just proceed with the next test.

Crossed-Polaroid Test

This test, which was described in the preceding chapter, is normally used to distinguish rubies and sapphires from stones like garnet, glass, and spinel. The results of this test can also be a warning sign of a synthetic. Look at figures 9.10 to 9.13 of the previous chapter. Notice that when the synthetic pink sapphire is in the table-down position, it changes from very dark to very light as it is turned 45 degrees between crossed-Polaroid filters.

If the stone were natural, it would tend to stay dark (in some cases it would stay light) as the stone is turned in the face-down position. Rainbow colors might also appear in the stone. This is because natural corundum is normally cut so it behaves like a singly refractive stone when viewed through the table facet. It's most intense color is normally visible when the stone is cut in this manner. (Technically stated, natural corundum is normally cut so its optic axis, the direction of single refraction, is perpendicular to the table).

A cutter may orient the natural ruby or sapphire differently if he wants to lighten its color, reduce the visibility of flaws, or make its color look even. Nevertheless, natural corundum normally does not show a distinct change from dark to light when viewed through crossed Polaroids in the table-down position, especially if it has a uniform color.

If evenly colored corundum changes from light to dark in the table down position, it's likely to be a synthetic. Synthetic corundum is typically cut in a way that produces this reaction.

Two-Color Test

The two-color test described in the preceding chapter can help separate natural from synthetic corundum like the crossed-Polaroid test. Place the perpendicular Polaroid strips in front of the table of the stone and try to find the two dichroic colors--for example, violetish blue and greenish blue in the case of a blue sapphire. If both colors can be seen through the table with the Polaroid strips, it is most likely a synthetic. Synthetic corundum is often cut with these two colors showing through the table. Natural corundum that is evenly colored normally isn't cut this way.

Bubbles Test

Synthetic stones frequently have tiny gas bubbles that may be round, tadpole shaped, or cocoon shaped. In the newer synthetics, it's not easy for a layperson to find these. Also, specks of dust or minute crystals may be mistaken for bubbles. Occasionally, though, there are bubbles that are obvious as in figure 10.2. When present in corundum, these bubbles indicate that the stone is a synthetic.

Fig. 10.2 A cloud of tiny gas bubbles in synthetic ruby

Other Tests

Jewelers and gemologists normally use a combination of tests to identify synthetics, and many of these tests require technical equipment and lots of experience. If you're interested in learning other ways in which synthetics are identified, some are listed briefly below:

- High magnification through a microscope is used to study the various kinds of inclusions. Certain ones indicate the stone is natural--for example, silk or zircon crystals with halos around them. Others prove it's synthetic.

- A light is directed through the stone with an instrument called a spectroscope to measure how it absorbs light. This test is particularly helpful in separating blue, yellow, orange, and green sapphires from their natural counterparts.

- The stones are placed under ultra-violet light and their fluorescence and transparency is compared to that of natural stones. Fluorescent reactions of natural and synthetic stones are listed in the appendix.

- The stones are exposed to x-rays, and then their fluorescence and phosphorescence is analyzed.

Types of Synthetic Ruby and Sapphire

Not all synthetics are created equal. Some can be made in less than a day at a cost of a few cents per carat. Others take up to a year to grow and may sell for several hundred dollars per carat. The three main types of synthetic corundum are described below.

Flame fusion (Verneuil). This is the most common and least expensive type of synthetic corundum. It's made by melting powdered chemicals with a gas flame and then allowing the molten chemicals to cool and crystallize at normal pressure. Due to the way it forms, flame-fusion corundum tends to have curved growth bands or lines (fig. 10.1).

Some flame-fusion stones sell for less than $1 a carat. Others sell for more depending on their quality and especially on the cost of having them cut. High quality cutting can add a great deal to their cost.

Melt pulled (Czochralski). This type tends to be flawless and is mostly used in industry for lasers, watch crystals, and optical instruments. It's produced by lowering a corundum crystal to the surface of molten chemicals. They solidify around the crystal as it is gradually pulled back up. Melt-pulled corundum can grow in a few hours, but costs more to produce than the flame-fusion type. A limited amount of it is used in jewelry, some of which is sold under the trade name "Inamori."

Flux growth. This is the most expensive type of synthetic corundum and the one which can be most easily mistaken for the natural stone. Its inclusions can look very similar to those in natural corundum.

Flux-growth ruby and sapphire is made by dissolving AL_2O_3 plus a coloring agent in a molten mixture. Then, as the mixture gradually cools or evaporates, crystals form. It may take from six weeks to over a year for flux-growth crystals to reach their desired size. Due to the longer time and amount of energy needed to grow them, these stones sell for a lot more than flame-fusion corundum--up to several hundred dollars per carat.

Compared to flame-fusion corundum, the production of the flux-growth type is relatively small. Most of the flux-growth corundum on the market has been produced for jewelry under the trade names "Chatham," "Ramaura," "Kashan," "Knischka," and "Lechleitner."

Synthetic Versus Natural

While reading this chapter, some people may understandably ask, "Why pay a fortune for a natural ruby or sapphire, when you can get a synthetic one for so much less?" This is almost like asking, "Why buy an original Rembrandt or Picasso painting when you can get a reproduction of it for much less." One of the main reasons is that it can be a pleasure to own a product of the original creator.

Another reason for buying natural gems is that they may possibly appreciate in value, whereas synthetics never will. There's a limited supply of good-quality natural rubies, for example. Lab-grown rubies, however, can be produced in whatever quantities are needed.

Some people may also be asking, "Why did scientists have to complicate the gem business by creating synthetics?" It's true that synthetic rubies and sapphires have been used to deceive people. Also, the process of identification has become more complicated, but there is a positive side too. Some of the reasons why we should be glad synthetics are available are:

♦ They cost a lot less. People that can't afford natural stones are still able to own a real ruby or sapphire.

♦ They benefit industry. Inexpensive synthetic corundum is used as an abrasive and is needed for bearings, lasers, and even ball-point pens.

♦ They have boosted consumer interest in natural corundum. As a result, natural stones have become more valuable, and stores are offering a wider selection of ruby and sapphire jewelry.

♦ They may have a more desirable color and clarity than the natural stones which are available. It can be hard to find high quality natural rubies and sapphires, especially in larger sizes. Synthetics offer jewelry buyers another option.

Quiz (Chapters 9 and 10)

Select the correct answer.

1. A two-carat, high-clarity, brilliant red stone that costs $7 in a retail store could be:
   a. Natural red spinel
   b. Red topaz
   c. Natural ruby
   d. Lab-grown ruby

2. Which of the following is **not** characteristic of glass stones?
   a. Gas bubbles
   b. Hexagonal bands of color
   c. Facets that curve slightly inward like a bowl
   d. Rounded facet edges

3. Closed-back settings:
   a. Are preferred by jewelers because they prevent the bottom of the stone from chipping.
   b Are preferred by jewelers because they prevent the bottom of the stone from getting dirty.
   c. Can be a sign that a stone is an imitation or foil backed.
   d. Allow the most light into a gemstone.

4. You place a red stone between two crossed polaroid filters. No matter how you position the stone, it stays dark as you turn it. The stone might be:
   a. Synthetic ruby
   b. Natural ruby
   c. Glass
   d. Padparadscha

5. Which of the following could help a gemologist distinguish a natural sapphire from one that is synthetic?
   a. Refractive index test
   b. Hardness test
   c. High magnification
   d. All of the above

6. Which of the following is **not** another name for synthetic ruby?
   a. Created ruby
   b. Cultured ruby
   c. Lab-grown ruby
   d. Imitation ruby

7. A large, blue, rectangular stone with only 5 facets on the crown is most likely
   a. Synthetic sapphire
   b. Natural sapphire
   c. Glass
   d. Blue topaz

8. Synthetic corundum is used for:
   a. Birthstone jewelry
   b. Ball-point pens
   c. Lasers
   d. All of the above

9. Which of the following is characteristic of synthetic ruby?
   a. Silk
   b. Curved lines that extend across the stone
   c. Concave facets
   d. Circular fractures around crystals

10. When you look at a red stone through two perpendicular polaroid strips or through a dichroscope, you see both orangy-red and purplish-red colors side by side. The stone might be:
    a. Synthetic ruby
    b. Glass
    c. Garnet
    d. Red spinel

True or False?

11. It's often easier to identify synthetic ruby than synthetic sapphire

12. A major advantage of buying antique jewelry is that you don't have to be concerned about the stones in it being lab-grown.

13. Jewelers with 20 to 30 years experience may accidentally sell a synthetic as a natural stone if they have not kept current on the development of new synthetics.

14. Synthetics often have a higher clarity than the natural stones they duplicate.

15. The inclusions of synthetic rubies and sapphires are always very different from those of natural stones.

16. Gemologists use a combination of tests to determine the identity of gems.

Answers

1. D - A top-quality, natural red spinel or red topaz is far more valuable than any laboratory-grown ruby, no matter what process was used to grow it. A red stone doesn't have to be a ruby to be valuable and desirable.

2. B

3. C

4. C

5. C - A synthetic sapphire has the same hardness and refractive index as a natural sapphire. In general, gemologists avoid hardness tests because they can damage the stone and are unnecessary.

6. D

7. C - Even large synthetic stones would normally have more than five facets on the crown.

8. D

9. B

10. A

11. True - The curved bands of synthetic sapphire can be very hard to see.

12. False - Lab-grown corundum has been used in jewelry since the early 1900's. It can also be found in jewelry with 18th- or 19th-century hallmarks. It's not unusual for stones of antique pieces to be lost or damaged and then replaced with synthetics.

13. True - That's why it's best to deal with knowledgeable jewelers who feel continuing education is important.

14. True

15. False - Flux-growth synthetics can have inclusions that look very much like those of natural stones.

16. True

11

Treatments

Imagine that you are at a beauty pageant judging the contestants. Would you disqualify them for wearing makeup, or would you expect them to have it on?

We may think it's fine for beauty contestants to enhance their appearance, but we're not always as open-minded when judging another product of nature--gemstones.

For centuries, rubies and sapphires have been heated to improve their color. However, in the past 20 years, heat treatment has been done on a wider scale and at much higher temperatures--1600°C (2900°F) and above. Thanks to this new technology, corundum previously considered unsaleable can now be cut into gems to meet increased world demand.

Heat treating is so common that the Jewelers of America association states in their consumer leaflets, "Virtually all rubies are heated to permanently improve their color and appearance," and "Virtually all blue, yellow and golden sapphires are heated to permanently produce or intensify their color."

Not everybody in the trade has been this candid about heat treatment. Some have felt uncomfortable about admitting it to the public. Consequently, heating gems has at times been viewed as something evil when in fact it's an ingenious technique for increasing their value and beauty. What is evil is to deny that these treatments are standard practice. Fortunately, the trade is becoming more open about discussing enhancements of all types.

Not all corundum treatments are regarded as equal. Even though it is considered normal to hide cracks in emeralds with oil, this practice is generally frowned upon when applied to rubies and sapphires. The next section will help you understand why some treatments are more accepted than others. It will also briefly outline what their purpose is and how they are detected.

Ruby & Sapphire Treatments

Heat Treatment

In the 1500's, heat was used to turn blue sapphires into colorless diamond imitations. Nowadays, it's used to turn off-white or near-colorless sapphires blue. Besides lightening and darkening sapphire, heat can also create a star in a stone or improve its clarity. Rubies are heated to make them appear less brownish or purplish and to improve their clarity.

Do not try to heat your rubies or sapphires in an oven to make them look better. They could crack, melt, explode, or turn colorless. Heat treating is a specialized skill that is best done by professionals. Finding competent heat treaters, however, can be difficult.

Heat treating is widely accepted because it causes a permanent improvement of the entire stone. Nevertheless, high-quality heat-treated stones are often valued less than their untreated counterparts. Untreated rubies and sapphires are rare, and rarity is prized in the jewelry trade. In commercial grades, it doesn't matter whether a stone was heated or not. The overall quality determines the price.

To detect heat treatment in rubies and sapphires, gemologists usually have to examine them under magnification. Heated stones may have fuzzy color areas and bands, surface pockmarks, melted facets, dot-like rutile "silk," or glassy circular cracks around crystals. Fluorescent reactions to ultraviolet light are also studied. Heat-treated blue sapphire, for example, often turns a faint chalky green under short-wave U.V. light.

It's not always possible to determine if a stone has been heat treated or not. However, when you shop for rubies and sapphires, you may as well assume that they have been heat treated. There are a few exceptions, four of which are:

- Rubies and sapphires that already have a high color and clarity when mined. They are not normally subjected to heating because this could damage them or remove their color.
- Some Tanzanian corundum which does not pass through Bangkok.
- Highly flawed stones that could be further damaged by heating.
- Sapphires found in the Yogo Gulch area of Montana, USA. Their natural color usually looks good. Plus, they don't respond well to heat treatment. Sapphires from other parts of Montana, however, are commonly treated.

Irradiation

Colorless sapphires from Sri Lanka are occasionally irradiated to make them yellow. Sri Lanka's pink sapphires may be irradiated to turn them into padparadschas. Irradiation also occurs naturally. Many gems have been colored by radioactive elements in the earth's crust.

Irradiation treatment of sapphire is not widely accepted because the resulting stones fade quickly in light and heat (irradiated gems like blue topaz, however, do not fade). Another problem is that the irradiation is sometimes done improperly, and the stones may end up being radioactive. Dangerous material is rare, but to be safe, reputable corundum suppliers check irradiated yellow sapphire and padparadscha with a radiation detector.

Irradiated sapphire is not a great concern. Most of the yellow sapphire on the market has not been irradiated and does not fade.

Oiling and dyeing

Low-quality rubies or sapphires (particularly cabochons and Indian star rubies) are sometimes dyed with a colored oil to hide cracks and improve color. To be oiled, stones must have surface cracks which allow the oil to penetrate. Therefore, oiling is usually reserved for low-quality, flawed corundum.

Unlike emeralds, faceted rubies and sapphires normally do not have a lot of surface cracks. Consequently, it is not common to oil them. However, if you travel to Thailand, you may see bottles of red ruby oil for sale. It is interesting to read the labels, especially if they are written in broken English. An example of one such label is: "*Crown Rubies Red Star*--Perfectly increasing 100% value of gems, rubies and sapphires with very shining and brightness. Soak your precious stones in the solution *Crown Rubies Red Star* as long as required, then clean and polish with cloth." This ruby oil, however, is generally used to add sparkle and shine to ruby rough rather than to faceted stones. Since the oiling of corundum is not permanent nor usually necessary, it isn't very well accepted by the trade.

If you see a lot of surface cracks in a ruby or sapphire, ask the store to clean the jewelry piece in their ultrasonic cleaner for a few minutes. Then check if the cracks are more visible or if the color has changed. If the store feels the stone is too flawed to be safely cleaned in an ultrasonic, this too indicates you would be better off buying another stone. (Do not ask to have emeralds cleaned in ultrasonics. Assume they have been oiled.) If you already own a ruby or sapphire with lots of cracks, do not do these tests. Take extra precautions when cleaning the stones (just use a mild soap and warm water), and do not put the stones in ultrasonics. Stones can be reoiled if the oil dries up or discolors, but this treatment should be done by a professional.

There is nothing wrong with oiled and dyed stones as long as the customer is informed that they are treated and told how to care for them. Oil and dye treatments provide a practical means of making low-grade corundum look better. People who otherwise couldn't afford a natural ruby are able to buy one that looks acceptable. Unfortunately, oiled and dyed stones are often sold with the intent of fooling buyers. Then, instead of being legitimate treatments, oiling and dyeing become deceptive practices.

Surface diffusion

This treatment is usually done to turn pale or colorless sapphires blue. It may also be used to turn stones red, orange, or yellow or to form a star. The pale stones are packed in chemical powders and then heated to 1700°C and above until a thin layer of color covers their surface.

Surface diffusion is relatively new (about 15 years old according to patent records) and is not very well accepted by the trade yet. It is becoming more prevalent, though, and is used on sapphire that does not respond to standard heat treatment. The color is permanent, but is only on the surface of the stone. Consequently, the color can be polished or scraped off leaving the grey interior exposed. Some people sell diffusion-treated stones openly, but others try to pass them off as nontreated.

Gemologists detect surface diffusion by immersing stones in water or glycerine or methylene iodide. Stones that are diffusion-treated will show some of the following characteristics: strong concentrations of color along cracks, facet edges, or the girdle; colorless areas; a blotchy color; and a high relief (untreated stones tend to fade into the background).

Disclosing Treatments

There is a great deal of controversy in the trade about the disclosure of gem treatments. Some feel that disclosure will confuse customers and hurt the colored-stone business. Others feel that disclosure is unimportant because treatments like dyeing, oiling, and heating have been practiced for centuries. Some describe treated gems as "faked" or "adulterated" and prefer to believe that the stones they sell are not treated when in fact they often are.

During the past ten years, however, the trade has become more aware of gem treatments and the trend now is towards full and voluntary disclosure of them to the public. Nevertheless, many jewelry salespeople admit they only discuss gem treatments when they are specifically asked about them. They offer some valid reasons for this.

Some say that the majority of their customers are only interested in finding a pretty style at a low price with easy credit terms. These salespeople feel it is pointless to discuss heat treatment with customers who are not even interested in knowing the quality or carat weight of the stone(s) they are buying.

Some say that when they have brought up the subject of treatments, they have had customers interrupt and say "Please don't tell me about that. I want to believe my stone is completely natural, color and all."

Others are very frank and claim that if they discuss treatments with customers, "it might kill the sale."

In spite of all this, consumers do have a right to know what they are buying, especially when they are spending large sums of money and when they show an interest in jewelry quality. If you wonder whether a ruby or sapphire you are buying is treated in any way, don't wait for the salesperson to tell you, ask. (However, they may not be able to give you an answer. Either the salesperson may not be informed about treatments or there may not be enough evidence to determine if the stone has been treated or not.) If you are spending thousands of dollars, it is advisable to get a report on the stone from a reputable gem laboratory. It would be important to know, for example, that a high-quality 7-carat sapphire is not heat treated. Due to its rarity, the stone could be worth 25% to 40% more than if it were treated (The price of small and average- or low-quality stones, however, normally is not affected by evidence of heat treatment). Likewise, it would be important to know that a large, high-quality yellow sapphire has a stable color instead of one that could fade due to irradiation treatment. Be sure to specify that you want the laboratory to indicate if there is evidence of treatments. Some labs only include this information when asked.

Before we become overly critical about jewelry professionals who don't voluntarily disclose treatments, we should probably take a look at the ethics of members of other professions. How many of our nations' leaders, for example, have been realistic about their campaign promises and have voluntarily admitted their weaknesses?

How many insurance agents voluntarily tell us what their policies do not cover. There are some, and they are the ones we are most likely to respect and trust.

Likewise there are some jewelers that present all the facts up front. They point out both the good and bad quality aspects of their goods, and they disclose treatments voluntarily. This indicates that they care about you and that they operate under a higher set of ethical standards than many of their competitors.

Thanks to these jewelers and other members of the trade, we now have laws requiring voluntary disclosure of treatments. Ethical jewelers have always pushed for laws against fraud. They deserve our respect and support.

12

Deceptive Practices

There is no fraud or deceit in the world which yields greater gain and profit than that of counterfeiting gems. (1st century AD, Pliny the Elder, Roman scholar, from his *37th Book of the Historie of the World*)

The counterfeiting of gems is as widespread as ever. There are just more ways of doing it now. In Pliny's day, there were no lab-grown rubies or sapphires. Irradiation and diffusion treatments were unknown. There's nothing wrong with creating synthetics or treating gems. Fraud occurs when a customer is not told that his stone is synthetic or treated. Fraud also occurs if a customer is misled into believing that a lab-grown ruby, for example, is as valuable as if it were natural.

Listed below are practices that are normally done with the intent to deceive. All of them, however, can be considered legitimate when they are properly disclosed to buyers.

Filling cavities (holes) with glass or another substance

It's not uncommon for a ruby or sapphire to have pits or cavities, especially on its pavilion (bottom). Gem cutters intentionally leave them on the stone to avoid trimming away valuable gem material. In the early 1980's, rubies with glass-filled cavities began to appear on the market. This is a clever way of filling holes and producing a smooth surface, but it creates some problems.

Glass fillings add weight to the stone, and it is hard to determine what percentage of the weight is not corundum. The glass fillings are softer and less durable than the rest of the stone and as a result are susceptible to breakage and chipping. Stones with glass fillings cannot be safely cleaned in ultrasonic cleaners. Ultrasonic cleaners not only shake the dirt loose. They may also shake out the fillings.

Gemologists detect glass fillings by reflecting light off the stone. Then they examine its surface under magnification for areas of different luster. They may also immerse it in a methylene iodide solution to spot the filling. If you notice areas of different luster on a stone, you shouldn't automatically assume they are glass fillings. They could also be naturally occurring glass or mineral inclusions. So have the stone checked by a professional before making a judgement.

Glass fillings are not common. In fact many jewelers have never had an opportunity to see them. Shellac and epoxy fillings in cabochon stones are more frequent. Black star sapphires, in particular, often have their pits filled with shellac.

Quench Crackling

Stones that are quench crackled have been heated and then plunged into cold water. This procedure is done to produce cracks in synthetic stones so they'll look more natural. Sometimes oil or other liquids are forced into the cracks to imitate the fingerprint inclusions found in natural corundum.

Foil Backing

Foil backings have been used to add color and brilliance to gems for probably 4000 years. As gem-cutting techniques progressed and brought out more brilliance in stones, these backings became less popular. Today foil backings are occasionally found on corundum cabochons, but they are more likely to be seen on glass imitations of ruby and sapphire (fig. 12.1). (See figures 9.2 and 9.3 in Chapter 9 for examples of closed-back settings). Antique jewelry buyers should be especially alert to the possibility of foil backings since they used to be very common. Beware of closed-back settings. Something such as foil may be concealed, particularly if a stone is unusually bright.

Fig. 12.1 A foil-backed glass stone

Composite Stones (Assembled Stones)

Composite corundum stones (those formed from two or more parts) are not very common. This is due to the availability of inexpensive synthetics, which are easier to manufacture. Nevertheless, we need to be aware that they are occasionally used to trick buyers.

The key to identifying a composite stone is to find where its parts have been joined together. This can often be seen by immersing the stone in rubbing alcohol or a methylene iodide solution (immersion tends to make color differences and the glue layer more obvious). Magnification is also helpful. It can reveal separation lines, flattened air bubbles between the parts, or swirly areas where the stone has been brushed with glue.

Stones composed of two parts are called **doublets**. Those consisting of three parts are called **triplets**. Listed below are various types of doublets or triplets that are used as corundum substitutes:

♦ **Natural corundum + natural corundum**. Natural corundum stones may be glued together for a couple of reasons. One large stone (especially if it's over 1 carat) can be sold for a higher per carat price than two smaller ones. Also, composite stones may have a more valuable color than their individual parts. For example, pale yellow sapphire pieces may be cemented with a blue glue to form a blue sapphire.

♦ **Natural corundum + synthetic corundum**. This is one of the most common types of composite stones. When examined under magnification, it may appear completely natural due to the presence of natural inclusions. The pavilion is usually synthetic red or blue corundum, and the crown consists of natural ruby or else natural sapphire that is either green, pale yellow, or light blue. The resulting stone is either red or blue, depending on the color of the pavilion.

♦ **Natural corundum + imitation**. This type of composite stone is rare. One example is natural white or grayish star sapphire capped with transparent red plastic to look like star ruby.

♦ **Synthetic corundum + imitation.** It's hard to imagine why anyone would bother making a stone from synthetic corundum and imitation material like synthetic spinel, but occasionally it is done. In fact, according to a report in *Gems and Gemology*, a ring set with a large synthetic spinel & synthetic ruby doublet and several diamonds was sent to the New York GIA Gem Trade Lab for identification.

♦ **Imitation + Imitation**. The best known imitation corundum doublet is the garnet and glass doublet. It was invented in the mid 1800's to imitate gems of every color. It's a more suitable ruby and sapphire imitation than glass because the garnet crown is more durable and adds luster to the stone. If you own or have an interest in antique jewelry, you should be especially aware of these doublets. A lot of the red and blue stones in expensive looking antique pieces (especially those of the latter half of the 19th century) are nothing but garnet and glass doublets. Now lab-grown rubies and sapphires are being used instead of these doublets.

Misnomers

Sometimes gems are sold under misleading names. A garnet, for example, may be called a California ruby to make it seem more valuable. If a salesperson adds a qualifying word to a gem name, ask him to explain what it means. Some misnomers for ruby and sapphire are as follows:

American ruby	garnet
Australian ruby	garnet
Balas ruby	spinel
Bohemian ruby	rose quartz
Brazilian ruby	topaz
California ruby	garnet
Cape ruby	garnet
Colorado ruby	garnet
Montana ruby	garnet
Siberian ruby	tourmaline
Spinel ruby	spinel
Brazilian sapphire	tourmaline or topaz
Spinel sapphire	spinel
Water sapphire	iolite

Quiz (Chapters 11 and 12)

Select the correct answer.

1. The owner of a jewelry store tells you that he only sells untreated stones. His ruby, sapphire, and emerald jewelry ranges in price from $50 to $2500. You should assume that:

   a. The stones in the store are all of inferior quality because they have not been treated to improve their appearance.
   b. This store is probably one of the best places in town to buy jewelry.
   c. The stones in the store are either from Tanzania or Yogo Gulch, Montana.
   d. The jeweler is probably either misinformed about gem treatments or else he is intentionally misrepresenting his merchandise.

2. Which of the following can be safely cleaned in an ultrasonic cleaner?

   a. An expensive, near flawless, faceted blue sapphire.
   b. A ruby cabochon with a lot of surface cracks.
   c. An emerald.
   d. A ruby with a glass filling.

3. The heat treatment of rubies and sapphires can:

   a. Improve their color
   b. Improve their clarity
   c. Create a star effect
   d. All of the above

4. Which of the following treatments is <u>**not**</u> used on rubies?

   a. Heat treatment
   b. Irradiation
   c. Glass fillings
   d. Dyeing

5. Treatments:

   a. Always lower the value of a gemstone.
   b. Always increase the value of a gemstone.
   c. Allow consumers to have a better selection of rubies and sapphires.
   d. Have only been used on gemstones during the past 200 years.

6. Closed-back settings may be a sign that a stone might be:

   a. foil backed
   b. a composite stone
   c. either of the above
   d. neither of the above

Match the following "selling terms" (euphemisms) to the correct equivalent term.

7. Enhancement
8. Identifying characteristic
9. Created ruby
10. Feather

a. Flaw
b. Synthetic ruby
c. Crack
d. Treatment

Answers

1. D - Even though there are a few dealers that specialize in untreated gems, the price range of their goods would not be between $50 and $2500, and the stones would probably be unset rather than mounted in jewelry. Emeralds in this price range are likely to be oiled no matter where they are from.

2. A

3. D

4. B

5. C

6. C

7. D Enhancement = treatment

8. A Identifying characteristic = flaw

9. B Created ruby = synthetic ruby

10. C Feather = crack

13

Star Rubies & Sapphires

Mr. Johnson was in Bangkok for a convention. While he was there, he wanted to get a star sapphire--preferably one with a strong blue color and a distinct star having long straight rays. He had heard this was the best kind.

Even though he went to several stores, Mr. Johnson didn't seem to have much luck finding such a stone. Most of thc star sapphires he saw were either pale or white or gray. It seemed that the bluer the stones were the more blurry or imperfect the star was. Finally, he spotted a star sapphire that looked even better than one he had seen at the Smithsonian museum. To his dismay, it was a lab-grown stone.

When buying gems, we have to be realistic about our expectations. We must be aware, for example, that natural emeralds normally have cracks. Likewise, we need to know that natural star sapphires are normally more pale than natural faceted sapphires, and their stars are not as well defined as those of laboratory-grown (synthetic) stones.

The first synthetic star corundum stones were produced in 1947 by the Linde Division of the Union Carbide Corporation in the United States. Ever since they were introduced to the jewelry trade, there has been a tendency to expect natural stones to resemble them. It's true that there are some very fine deep blue and red specimens in museums and private collections, but these are the exception rather than the rule.

Even though lab-grown star sapphires and rubies usually have sharper stars and a more intense color than a natural stone, they are not highly valued. They can be found in jewelry supply stores for between $10 and $30. (However, some of the newer synthetic stones with lighter colors and more natural looking stars sell for a lot more.) In contrast, a natural stone with a similar color and size but a less perfect star than a $20 synthetic could sell for several thousand dollars.

It has always been hard to find top quality star sapphires and rubies. It's now getting harder to find them because the heat treatment of corundum has become so widespread. The unusually high temperatures at which the stones are heated often melt the silky mineral fibers in them that are responsible for creating the star effect.

Two varieties of star corundum that are still relatively easy to find are Indian star rubies and black star sapphires from Thailand and Australia. Indian star rubies are usually nontransparent and their red or purplish hue is masked by a lot of gray or brown. As a consequence, many of these stones sell for only a few dollars a carat. Black star sapphires with a white star are also sold at very low prices. If they have a good yellow star and a large size they may sell for up to $100 a carat in the local Thai market. These 'golden-star' black star sapphires are seldom sold outside of Thailand.

Most star rubies and sapphires have 6-rayed stars, but occasionally their stars are 12-rayed. The extra rays appear when two different types of mineral fibers are present (rutile and hematite). Sometimes six of the rays are yellowish and the other six look white. These stones are rarely sold in jewelry stores. When available, they do add interest to jewelry pieces.

Star rubies and sapphires have often been worn as good-luck charms. In his book *The Curious Lore of Precious Stones*, George Kunz states that the three cross-bars of the star were thought to represent faith, hope, and destiny. Supposedly, star sapphires were so powerful at warding off bad omens they would exercise their good influence over their first owners even after passing into other hands.

Judging Quality

Color

The evaluation of color in star rubies and sapphires is similar to that of faceted stones, but the overall grading is more lenient. Generally, the more saturated and pure the body color, the more valuable the stone is. Medium and medium-dark tones tend to be the most prized. Light tones, however, are considered acceptable.

Red is regarded as the most valuable hue with blue being second. Black star sapphires and purplish or maroon Indian star rubies are the lowest priced. As an added note, purplish and pinkish star corundum is often called star ruby, even in North America and Europe.

As mentioned earlier, the color of the star itself may also affect the price. Yellow-star black sapphires are more valuable than those with white stars.

The Star

Your main concern when judging a star ruby or sapphire should be: Is it easy to see the star when you look at the stone under a single source of direct light? Some secondary questions to ask are as follows:

- Is the star centered?
- Is the star sharp and well defined?
- Are the rays straight?
- Are all the rays present?
- Do the rays extend completely across the stone?
- Is there a good contrast between the star and the background?

Ideally you should be able to answer yes to all of the above questions. In actuality, though, the stars on natural stones tend to be slightly wavy, a little blurry, and/or incomplete. Often the better the color is, the more imperfect the star looks. Appraisers normally indicate the degree to which the stars conform to the above standards. Under the category of star centering, for example, they may indicate poor, fair, good, very good or excellent.

Transparency

The degree of transparency plays a major role in determining the value of star corundum. But sometimes this factor is overlooked. Gemologists use the following terms to describe gem transparency (the amount of light which passes through the gem):

- **Transparent**--objects seen through the stone look clear and distinct.
- **Semitransparent**--objects look slightly hazy or blurry through the stone.
- **Translucent**--objects are vague and hard to see. Imagining what it is like to read print through frosted glass will help you understand the concept of translucency.
- **Semitranslucent**--only a small fraction of light passes through the stone, mainly around the edges.
- **Opaque**--virtually no light passes through the stone.

The highest quality star rubies and sapphires are semitransparent. As a general rule, the more transparent a star stone is the greater its value. A translucent Indian star ruby, for example, may sell for 10 times more than if it were opaque. Most Indian star rubies tend to be opaque. Consequently, they usually cost less than white or gray translucent star sapphires.

Clarity

It is normal for star rubies or sapphires to have flaws. However, the more obvious these flaws are to the naked eye, the lower the value of the stone. It's usually best to avoid stones with a lot of surface cracks because they may not be very durable, and they may be dyed.

Be particularly careful with black star sapphires when you wear them or clean them (avoid putting them in ultrasonics). Black star sapphires have a tendency to split or crack or chip. In fact, a fair number of black star sapphires cut in Thailand have chips or holes that are filled with shellac. Therefore, it's a good idea to examine them under magnification before buying them.

Fig. 13.1 An Indian star ruby. Note the hexagonal patterns, typical of natural star corundum.

Fig. 13.2 A 12-ray star that is off-center

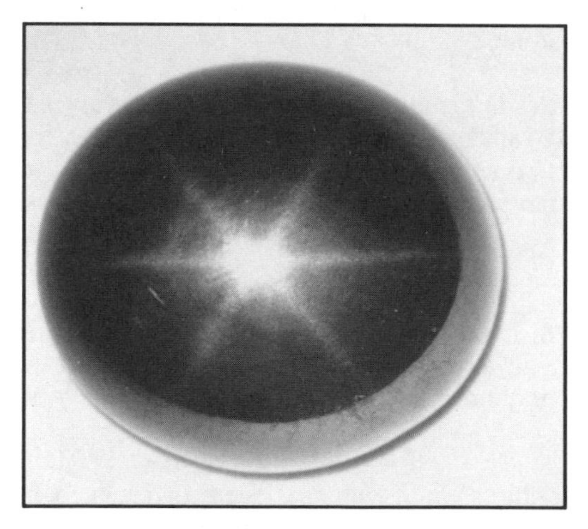

Fig. 13.3 A slightly blurry star

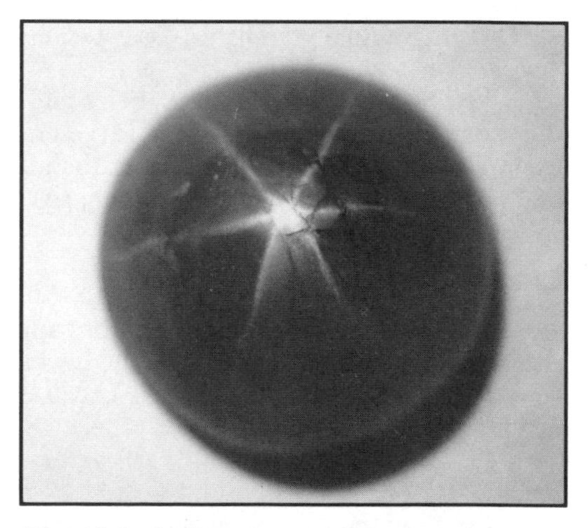

Fig. 13.4 A sharper star, but note the incomplete rays

Cut

When judging the cut, you must take into consideration the color and transparency of the stone. For example, you should be more strict with star stones that are opaque. They should have as little weight as possible below the girdle (preferably less than 1/4 of the total weight). Otherwise, you end up paying for a lot of weight that adds nothing to the beauty of the stone. Also, if the bottom is too deep, the stone may not be suitable for mounting in jewelry. Figure 8.3 of Chapter 8 is a good example of an Indian star ruby with too much weight on the bottom. Semitransparent stones, however, require a greater depth below the girdle to intensify their color and emphasize their star.

The proper height of a star stone also varies according to what type it is. High quality stones from Burma or Sri Lanka normally have a medium to high dome with a uniform curvature to create a good star effect. Black star sapphires, on the other hand, tend to be flatter because low domes make their stars stronger and sharper.

If you are fortunate enough to find a natural star ruby or sapphire with a good strong color, a distinct star, and a fair degree of transparency, there is no point in being overly concerned with how it is proportioned. If you have to pay for a lot of extra weight, just consider it as part of the price of owning a rare gem.

Genuine and Natural or Not?

Imitation and synthetic star rubies or sapphires are often fairly easy to spot. Some of the tests that will help you detect them are listed below:

The Perfect Star Test

Examine the star on the stone. If it looks too good to be true, you should be suspicious. The stars of most synthetic stones are sharper and straighter and more perfectly formed than those of natural stones. This is because the mineral fibers which produce the stars are much finer and more evenly distributed in synthetic rubies and sapphires. The stars of certain synthetic stones (the first ones produced or some of the latest ones) may look hazy and/or incomplete; but for the most part, the stars of lab-grown corundum tend to have a perfect "painted-on" look.

The Ray-count Test

Count the number of rays on the stone. Star rubies and sapphires should have either 6 or 12 rays. Sometimes black star diopside (another type of gem) is mistaken for black star sapphire. It's not hard to separate the two, though. Star diopside usually has 4 rays. Another stone that might have just 4 rays is star spinel.

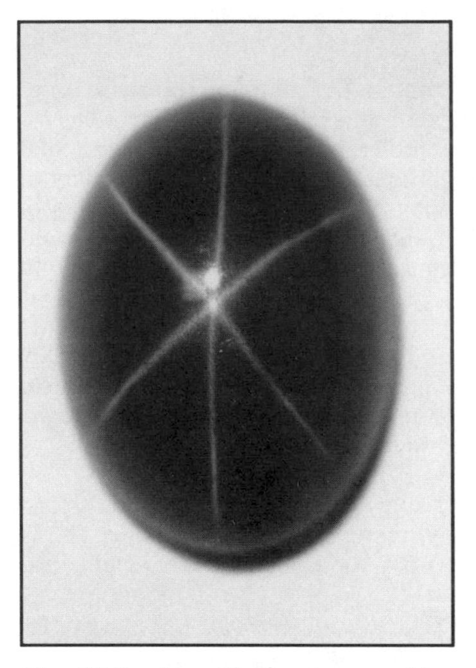

Fig. 13.5 A synthetic star sapphire. Note how long and sharp the rays of the star are.

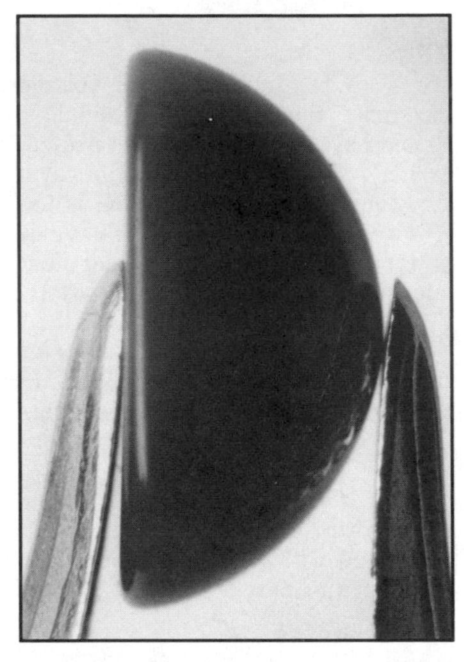

Fig. 13.6 Side view of a synthetic star sapphire with a flat base. Note how opaque and symmetrical it is.

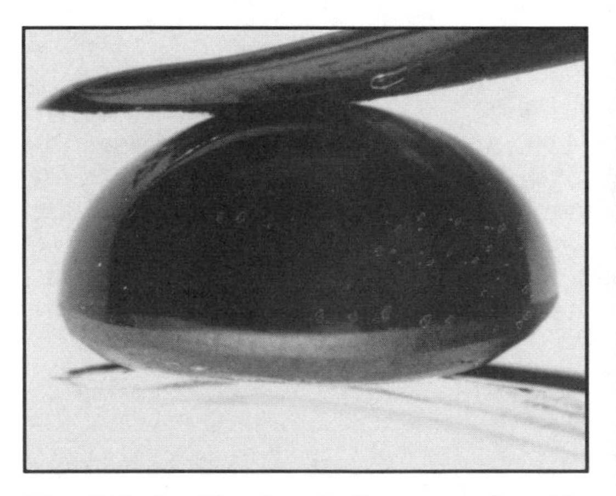

Fig. 13.7 Profile of an Indian star ruby with acceptable proportions

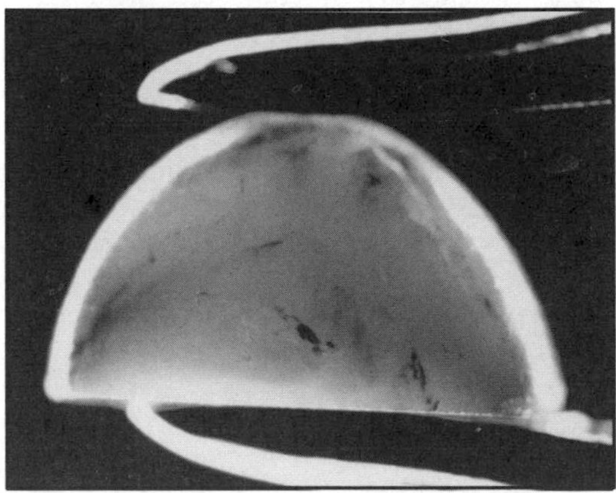

Fig. 13.8 Side view of a natural star sapphire with a flat base. Note the translucency and inclusions.

The Perfect Clarity Test

Look for flaws in the stone such as pits, cracks, hexagonal patterns, or colored specks and patches. These are common in natural stones. If you look at the stone through a 10-power magnifier, you may be able to see the fiber-like inclusions which create the star effect. If you do, this indicates a natural stone. Normally for synthetics, you need at least 50-power magnification to see their thin fibers.

The Flat-bottom Test

Look at the bottom of the stone. If it is flat, chances are the stone is synthetic. Natural star stones tend to have rounded bases. Occasionally, however, they are found with flat bottoms as in figure 13.8.

The Curved Band and Bubbles Test

Examine the stone under 10-power magnification for curved bands or bubbles. Their presence indicates the stone is synthetic. Bubbles are also commonly found in glass.

The Profile-view Test

Look at the stone from the side. It should normally look the same color as it does from the top of the stone. Occasionally, the stone may look almost colorless through the side. This occurs when a material such as star rose-quartz is backed with a colored foil to imitate star corundum.

Transparency Test

Is the stone opaque? If it is, and if the stone has a vivid red or blue color, chances are it's a synthetic. Lab-grown stones tend to be opaque or at best semitranslucent (but a few of the newer types are semitransparent). Natural star corundum with a fine blue or red color tends to transmit some light. Inexpensive black star sapphires and Indian star rubies, however, are typically opaque.

Closed-back Test

If the stone is set in jewelry, look at the back of the setting. Be suspicious if the entire bottom of the stone is blocked from view or enclosed in metal. Closed backs suggest that something is being hidden. Some examples of what might be concealed are listed below:

- Inscriptions indicating the manufacturer of a lab-grown stone, such as "L" for Linde.
- Colored foil backings used to give the stone a deep ruby or sapphire color.

- Fine lines engraved on synthetic corundum and spinel which create a star effect when viewed from above. The lines may also be engraved on metal or other material and then cemented to the bottom of the stone.

- A bottom rubbed with ordinary pencil lead to make the star look darker and more pronounced.

- Colored nail polish, paint, enamel, plastic, or another coating used on the base to improve the color.

- Separation lines which indicate that the stone consists of two or more parts. For example, natural corundum is sometimes joined to synthetic corundum to improve the color and produce a natural look.

Closed backs do not necessarily mean that a stone is an imitation or synthetic. However, if you are buying a "natural" star ruby or sapphire from someone you don't know, you would be better off selecting a stone that is loose or set in a mounting with an open back.

14

Geographic Sources

When dealers price a diamond, they're not concerned about where it was mined. They just determine the price according to the quality. Also, they all agree on what color is most valuable--D color, which is no color (fancy diamonds excepted.) The colored stone business is a different game. The place of origin can affect the price, particularly in high quality rubies and sapphires; and there is a difference of opinion as to what colors are most valuable.

Even though dealers agree that Burmese rubies are best, their concept of a good Burmese ruby varies. For example, Benjamin Zucker, a New York gem dealer and author, states in his book, *How to Buy & Sell Gems*, that the finest shade of Burmese ruby is a full-bodied red with a touch of orange in it. The GIA, in their *Gem Reference Guide,* identifies the finest quality Burmese rubies as being red to slightly purplish red with a medium-dark tone and vivid saturation. Richard Hughes, the executive vice president of the Asian Institute of Gemological Sciences in Bangkok, identifies two types of good quality Burmese rubies in his book *Corundum*--those which are slightly orangy red and those which are purplish red and have the strongest **red fluorescence** (a red glow when exposed to x-rays or ultraviolet rays such as those of the sun or an ultraviolet lamp). He goes on to say that ruby connoisseurs, however, prefer a slightly orangy-red ruby, of good clarity (eye-clean), and strong fluorescence.

The more one reads and the more dealers one talks to, the more one realizes that a "Burmese red" covers a range of hues, and it's debatable as to which is the best and most expensive hue. Dealers tend to charge the most for colors they and their clients prefer. This means that at the wholesale level, the best buys on orangy-red rubies can often be found from a dealer who prefers purplish-red stones and vice versa.

Factors other than hue must also be considered when describing a Burmese ruby. Fluorescence is one example. The next section outlines some of the various factors.

Describing Rubies and Sapphires by Place of Origin.

Place names like "Burma" are commonly used to describe corundum, but these terms can have different meanings. Burma (or Burmese) ruby, for example, can signify the following:

- Any ruby mined in Burma (Myanmar) irrespective of its quality.
- Any ruby that resembles a good quality ruby from Burma, no matter where it was mined.
- A ruby that has a report from a respected lab stating it was mined in Burma. Labs examine the inclusions and other characteristics of the stone to determine its place of origin.
- A typical high-quality ruby mined in Burma.

This section will focus on the last meaning by outlining characteristics of typical stones mined in Burma and other well-known corundum sources. When you shop for gems, though, remember that rubies and sapphires which are called "Burmese" or "Kashmir" do not necessarily come from these places. You need to ask salespeople exactly what they mean by these terms when they use them. If a ruby was actually mined in Burma, it will normally cost more than if it's from somewhere else, providing it's of good quality. If a respected lab identifies it as Burmese, the price will really go up. High-quality rubies and sapphires from Burma and Kashmir are extremely rare. Don't expect to find them at your local jewelry store.

(The following descriptions are based on sources listed in the bibliography as well as on discussions with dealers and jewelers in various areas of the world. They are only meant to be general guidelines, not absolute rules. Keep in mind that terms such as "red" and "dark" vary in meaning depending on the color-grading system used. Nevertheless, these terms can help us differentiate between the stones of various sources).

Characteristics of **high-quality Burmese rubies**:

- The hue ranges from purplish red to orangy red.
- The tone ranges from medium to medium dark.
- There is hardly any brown or gray masking the hue. (In other words, there is a minimal amount of brown or gray present. Consequently, the color of the stone is a more intense red.) The purer the color the better.
- They have a strong red fluorescence, an important but sometimes overlooked trait. This fluorescence can make purplish and orangy colors look red and helps mask dark extinction areas.
- Black extinction areas are at a minimum. The stones tend to look red throughout, even on the facets which are not directly exposed to light.
- The color is highly saturated as a result of the tone, purity of color, and strong red fluorescence.
- Pink overtones are often present.

Characteristics of **typical Sri Lankan (Ceylon) rubies**:

- The hue usually ranges from purplish red to red.
- The tone normally ranges from medium to very light. The lightest tones are the least valuable. In Europe and North America, medium light to very light red stones are called pink sapphires.
- They have a strong red fluorescence.
- There are relatively few dark extinction areas due to the red fluorescence and lighter color of these stones.
- The color is often unevenly distributed.

Characteristics of **typical Thai rubies**:

- The hue ranges from purplish red to orangy red.
- The tone usually ranges from medium dark to very dark, but the medium dark tones are more valuable.
- The hue is often masked by more gray, black, or brown than in Burma rubies. The purer the color the better.
- There tends to be very little or no red fluorescence. The more red fluorescence the better.
- There tend to be a lot of black extinction areas. The more red and the less black the better.

Characteristics of **typical Kenyan rubies**:

- The hue usually ranges from red to orangy red.
- The tone tends to range from medium to medium dark.
- They have a strong red fluorescence
- They tend to have a low transparency due to the presence of thick fingerprint and needle inclusions.
  Note: Better quality Kenyan rubies are often sold as Burmese rubies because of their similar color and fluorescence.

Characteristics of **high-quality Kashmir sapphires**. (Kashmir stones are the highest priced sapphires.)

- The hue ranges from violetish blue to blue.
- The tone ranges from medium to medium dark.
- Gray color masking is at a minimum. The purer the color the better.
- They have a powdery, velvety appearance.
- Dark extinction areas are at a minimum. Consequently, the blue is more predominant than in stones from other areas.
- Their color looks good in any type of light--daylight, incandescent, and fluorescent.
- The color is highly saturated due to the tone and color purity

Characteristics of **high-quality Burma sapphires.** (They follow Kashmir sapphires in terms of prestige.)

- The hue ranges from violetish blue to blue.
- The tone is usually in the medium-dark range, but some stones are lighter. For the most part, Burma sapphires tend to look darker than those from Kashmir and Sri Lanka.
- Gray color masking is at a minimum. The purer the color the better.
- The color tends to be more evenly distributed than in Kashmir and Sri Lankan sapphires.
- There tend to be more dark extinction areas than in Kashmir and Sri Lankan sapphires but fewer than in stones from Thailand and Australia.
- The color is highly saturated due to the tone, color purity, and uniform color.

Characteristics of **typical Sri Lankan sapphires.** (They follow Kashmir and Burma Sapphires in terms of prestige.)

- The hue ranges from violetish blue to blue
- The tone usually ranges from medium-dark to very light. The light and very light tones are the least valuable.
- The hue is often masked by gray. The less gray there is the better the color.
- The color is often unevenly distributed. The more even the color the better.
- Normally they have more brilliance and fewer dark extinction areas than other sapphires. This is mainly due to the lighter color of the Sri Lankan stones.
- Their color is usually less saturated than that of Kashmir and Burma sapphires. Light tones, gray color masking, and uneven color contribute to the lower saturation levels. The more saturated the color the greater the value.

Characteristics of **typical Thai & Australian sapphires.** (Thai sapphires have more prestige than Australian sapphires, but it's not easy to tell the difference between the two. Sapphires that are sold as Thai sapphires are frequently from Australia.)

- The hue ranges from violetish blue to greenish blue. Greenish-blue hues are generally considered the least valuable.
- The tone usually ranges from very dark to medium dark. The medium-dark tones are the most valuable.
- The hue is often masked by a fair amount of gray or black. The purer the color the better.
- There tends to be a lot of black extinction areas. The more blue and the less black the more valuable the stone.
- The color is less saturated than in Kashmir and Burma sapphires.

The sapphires in your local jewelry stores are most apt to be from Australia or Thailand. When you examine them, keep in mind that their quality is more important than their place of origin. It's mainly when high-quality stones are from Burma and Kashmir (and sometimes Sri Lanka) that country of origin affects the price.

Another thing to remember is that there are differences of opinion as to what are the best sapphires. Just as dealers disagree on whether orangy-red or purplish-red rubies are better, they also disagree on whether medium or medium-dark sapphires are better. In the United States, there is a tendency to prefer medium-dark tones, but in countries such as Japan and Germany, stones with medium tones and higher brilliance tend to be preferred. Prices vary according to the tastes of the dealers and their clientele. Some dealers say that at the wholesale level, the best buys on medium-dark sapphires tend to be from dealers who prefer medium tones (especially when they think the buyer prefers them too) and vice versa.

The ruby and sapphire sources listed above are among the best-known sources. However, they are not the only ones. Others will be briefly discussed in the next section.

World Sources of Rubies and Sapphires

Rubies and sapphires are found on every continent of the earth, but Asia has been the principle source of fine quality corundum. An alphabetical list of sources is given below. Also included is a brief description of the varieties of corundum found in each source. Table 14-1 summarizes where the various types of corundum are found. (Modern exploration is producing new developments. The status of the ruby and sapphire sources presented here will undoubtedly change in coming years.)

Afghanistan. Rubies with a color and fluorescence resembling Burmese and Sri Lankan stones are found in Jagdalek. However, they are usually very flawed. The political situation has restricted the output.

Australia. Queensland and New South Wales are major producers of blue sapphire, thanks to the development of heat treatment. This can make it more transparent and in some cases, lighten its dark color. Yellow and green sapphires and black star sapphires are also mined in significant quantities in these two states.

Sapphires were first discovered in Australia in 1851, but it wasn't until 1890 that commercial production began there. Most of the stones at the time were sold to Russia via German buyers. Some ruby has been found in the Harts Range of the Northern Territory, but it is too flawed to be of commercial importance.

Brazil. Both ruby and sapphire have been found in the states of Bahia, Ceara, and Goias, but generally their clarity is so low that they are only suitable for cabochons. Some green sapphire has been found in Bahia and sold under the name of "oriental emerald."

Burma (Myanmar). For the past 800 years, Burma has been renowned for producing the finest quality ruby (both faceted and star). Its blue sapphire, too, is highly prized. Yellow, pink, green, violet, and color-change sapphires are also found in Burma.

Since 1962, the country has been sealed off by the government, but mining in its principal gem tract at Mogok had been on the decline even before then. Today, fine Burmese gems have to be smuggled out of the country or else acquired through estate and second-hand sales.

Table 14.1 World Sources of Ruby & Sapphire

	Ruby	Blue Sapphire	Fancy Sapphire
Afghanistan	x		
Australia	x	major	major
Brazil	x	x	x
Burma	x	x	x
Burundi		x	
Cambodia	x	x	
China	x	x	
Colombia		x	x
India	x	x	
Kenya	x	x	x
Malawi	x	x	x
Madagascar	x		
Nepal	x		x
Nigeria		x	x
Pakistan	x	x	
Sri Lanka	x	major	major
Tanzania	x	x	major
Thailand	major	major	major
USA	x	x	x
Vietnam	x		
Zaire		x	
Zimbabwe		x	x

x = source, major = major source

Burundi. There have been recent reports of sapphire in this country, but little has been written about it.

Cambodia (Kampuchea). Cambodia is known for its ruby and blue sapphire from Pailin. The sapphires, which have a very pure and intense blue color, rank between those of Burma and Sri Lanka in terms of prestige.

Cambodian ruby is similar to that of Thailand and is produced in significant quantities. According to some reports, a large percentage of the ruby now cut in Thailand comes from Cambodia.

China. Gem-quality sapphires are found in two main areas of China--Hainan Island and Fujian province. Most of the rough sapphires are dark blue to greenish blue and resemble those from Australia and Thailand. Their average weight is about 2 carats. Low-quality ruby has been found in Sichuan Province.

Colombia. In the past few years, commercially important quantities of gem-quality sapphire have been found in Colombia. Most of the stones are blue, but they also come in a wide range of fancy colors--yellow, pink, lavender, and a slightly brownish green. Even though many of the stones tend to be pale, their colors can be intensified with heat treatment.

India India's claim to fame in the corundum industry is the Kashmir sapphire, considered the finest in the world. The production of sapphire began there about 1883, but most of it was over by 1937. There has been sporadic mining since then. Now, however, the Kashmir mines are officially closed.

India is also known for its ruby, especially its low-quality star ruby, which tends to be opaque and maroon in color. Most of the ruby material has come from Southern India. However, recently, Orissa in Eastern India has been producing fine-quality ruby (it's similar to Burmese) and mining there has been on the rise.

Kenya. Kenya is noted for its fine-color ruby. Most of it has only been suitable for cabochons, due to the presence of flaws. However, heat treatment has helped improve its transparency, allowing more of it to be faceted now.

In 1987, star sapphire from northwest Kenya began to appear on the market in commercial quantities. These stones often have sharp 6 or 12 ray stars and come in a variety of colors--blue, gray, brown, lavender, and greenish yellow. Large quantities of blue sapphire (similar to Australian) have also been discovered in the past few years.

Madagascar. There have been written reports of ruby from here as far back as 1922.

Malawi. This is a minor source of various colors of gem corundum--red, pink, blue, green, orange, and yellow.

Nepal. A small amount of ruby and pink & violet sapphire is produced here, but it tends to be heavily flawed.

Nigeria. During the past few years, Nigeria has become an important source of blue sapphire. The material is similar to that of Australia and Thailand. Green, yellow, and bicolored sapphire are also found here.

Pakistan Ruby and some purple to violet sapphires are found in the Hunza Valley. It is mostly of cabochon grade.

Sri Lanka (Ceylon). This is the most important source of fine quality corundum in the world in terms of quantity, variety, and size. Almost any color stone can be found here including the pink-orange padparadscha. Star and color-change stones are found as well. Unlike some sources which are relatively new, Sri Lanka has been an important producer of corundum for over 2000 years.

Tanzania With the exception of Sri Lanka, no other country produces as many color varieties of corundum as Tanzania. Very little of the material, however, is blue. The cut stones are usually below two carats and tend to have a lot of flaws. Recently, however, the clarity of these stones has been improving. Tanzania is particularly noted for its fancy-color sapphires as well as bi-color and change-of-color sapphires. Most of these come from the Umba Valley. Two main areas where rubies are found are Longido and Morogoro.

Thailand. Thailand is more than just a major producer of ruby and sapphire. It is the cutting and trading center of the world for corundum. Political events have contributed to Thailand's current stature in the corundum world. When the government of Burma changed in the early 1960's, corundum production there declined, and Burma ceased being the world's major supplier of ruby. Thai rubies had to be used to replace those from Burma.

Another reason why Thai rubies are more in demand now than previously is because heat treating processes have been developed to improve their color. Today, the majority of the rubies sold are of Thai origin (although it is hard to estimate what percentage might now be coming from Cambodia).

Thailand is also a major supplier of blue sapphire and produces significant quantities of black star and yellow sapphire as well.

United States. Montana, the most important corundum region of the US, is particularly noted for its Yogo Gulch blue sapphires. They have a very high clarity and a consistent, even-blue color. They do not need to be heat treated to look good, unlike sapphires from most other areas. Richard Hughes, in his book *Corundum*, states that if it were not for their small size (usually under 1 carat in the rough), Yogo sapphires might be the world's most renowned blue sapphires. The mine's most prolific production period was between 1896 and 1929. There are still significant quantities of sapphire at Yogo Gulch, but the cost of mining there compared to other countries of the world is too high to make it profitable. Today, Yogo sapphires are advertised in lapidary magazines and sold mostly to hobbyists, a contrast from 90 years ago when they were promoted in the Tiffany & Co. catalogue. A wide variety of light-color fancy sapphire (pink, yellow, green, lavender, etc.) have been found in areas west of Yogo, and production there is increasing. Montana might soon be a major supplier of fancy sapphire.

Photo Robert Weldon, GIA

1.02-ct Burmese ruby

Photo Robert Weldon, GIA

6.68-ct Kashmir sapphire, Hixon collection, Los Angeles County Museum of Natural History

Photo Tino Hammid, GIA

Ruby Carving

Photo Smithsonian Institution

138.7-ct Rosser Reeves Star Ruby

Photo Robert Weldon, GIA

Sapphire crystals from Sri Lanka

Photo Robert Weldon, GIA

Left--natural ruby, right--synthetic ruby

Photo Shane McClure, GIA

Pale-colored sapphires before irradiation treatment

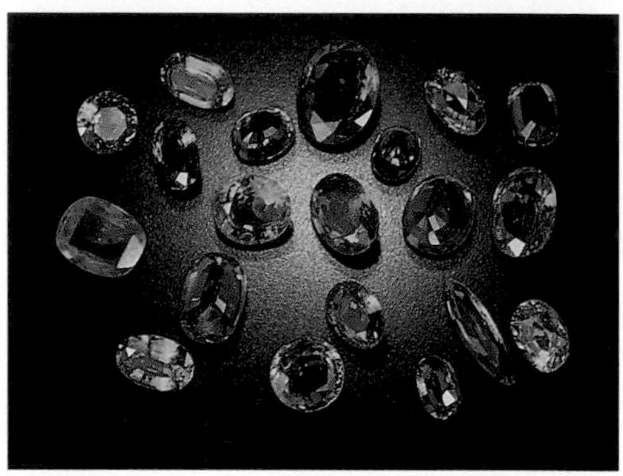

Photo Shane McClure, GIA

Same stones after irradiation treatment

Photo GIA

Natural padparadscha

Photo GIA

Tanzanian sapphires

North Carolina is the second most important corundum region of the US. Some ruby has been found there along with small amounts of sapphires in a variety of colors.

Vietnam. Ruby rough similar to that of Pakistan and Afghanistan has recently been found in Vietnam. It tends to be milky and very flawed. However, if it is heat treated, it might be made to look like material from Burma or East Africa.

Zaire. Blue sapphire, which tends to be grayish but transparent, is found in the Kivu region.

Zimbabwe (Rhodesia) High-quality blue sapphire (sometimes resembling Burmese material) has been produced here since 1968. Black star sapphire is found as well.

15

Choosing a Qualified Appraiser

Joe wanted to get his wife a sapphire ring for Christmas. There was one for $9000 that he liked; but if he bought it, he wanted someone not affiliated with the jewelry store to appraise it. The ring came with a 100% money-back guarantee, so he went ahead and got it.

When Joe started looking for an appraiser, he discovered it was not easy to find one. He looked in his local phone book (it covered a suburban section of Los Angeles), and the only jewelry appraiser listed under "Appraisers" was a coin shop. Fortunately, his neighbor had a phone book of another area of Los Angeles which had a few listings for jewelry appraisers.

Joe called them to find out about their qualifications and appraisal fees. One said she was a graduate gemologist. Another said he was a master gemologist appraiser. Another said he had been in the business for over 25 years. When Joe asked if he had any credentials, the man replied, "In the jewelry trade, it's experience that counts, not credentials." Two other appraisers described their experience and listed associations or societies they belonged to.

Joe didn't know which appraiser to choose. In fact, he had never heard of the credentials or associations they mentioned, and he wondered if you can learn much about an appraiser from a phone call.

Fortunately, you can learn a great deal about an appraiser over the phone. This chapter will help you interpret what appraisers tell you about themselves. It will then be easier for you to evaluate them.

Four questions you should consider asking when interviewing an appraiser are:

- What are your qualifications?
- How much do you charge?
- What does your appraisal fee include?
- Can you appraise the piece while I wait?

Some answers to these questions are discussed in the following sections.

Qualifications to Look For

To appraise a ruby or a sapphire, you need to know that it's not an imitation or a synthetic; and you must determine if it has been treated in any way. If "appraisers" tell you that "book-learning" and credentials are unimportant, they are implying that identification skills are also unimportant to them. Chances are they could be easily fooled by a synthetic stone. Competent professional appraisers should have one of the following diplomas to indicate they have basic gem identification skills:

- **FGA**, Fellow of the Gemmological Association of Great Britain
- **GG**, Graduate Gemologist (Awarded by the Gemological Institute of America)
- A gemologist diploma from another school or association

The FGA and GG are known internationally. No matter where you travel in the world, you will be able to find jewelry professionals holding these diplomas.

Although the above gemologist diplomas are important, they do not qualify a person to be an appraiser. One must also be familiar with gem prices, jewelry manufacturing costs, valuation theory, and the legal aspects of appraising. One must have trade experience, integrity, and the initiative to keep up with the market and new developments in appraising and gemology.

Consequently, good appraisers should be able to say a lot more about themselves than just "I'm a graduate gemologist." When asked about their qualifications, a competent appraiser might say they are or have been:

- An **MGA**, a master gemologist appraiser. This is the highest award offered by the American Society of Appraisers. To receive it, a person must pass their appraisal tests and have a gemologist diploma, an accredited gem lab, and at least 3 to 5 years appraisal experience.
- A **CGA**, a certified gemologist appraiser. This is awarded by the American Gem Society to certified gemologists that pass their written and practical appraisal exam. Trade experience is a prerequisite.
- A **CAPP**, a certified appraiser of personal property. This is the highest award offered by the International Society of Appraisers. To receive it one must attend their appraisal courses and pass exams. Trade experience is a prerequisite.
- An appraiser for jewelers. Then they may give you the names of the jewelers that use their services.

- An appraiser for banks, auction houses, and/or the government. Then they may give you references.

- A teacher of appraisal and/or gemology classes. Then they may tell you the names of the schools, organizations, or companies they are employed by.

- An author of appraising, jewelry, and/or gemology literature. Then they may list the publications they have appeared in or books they have written.

- Employed by wholesalers, retailers, gem labs, and/or jewelry manufacturing firms. Then they may describe their experience and/or tell you who they have worked for.

- A museum curator. Then they may tell you the name of the museum.

- An officer or member of appraisal and gemological associations. Examples in North America are:

  AAA, Appraisers Association of America
  AGA, Accredited Gemologists Association
  ASA, American Society of Appraisers
  CGA, Canadian Gemmological Association
  ISA, International Society of Appraisers
  NAJA, The National Association of Jewelry Appraisers

As you can see, there are lots of possible responses to the question, "What are your qualifications?" None of these responses by themselves, however, is a guarantee that an appraiser is ethical and capable of doing a high quality appraisal of an item such as a $9000 sapphire ring. This judgement should be based on a combination of factors, including the appraiser's general attitude when he or she talks with you.

Sometimes it's assumed that if a person has been in the jewelry business for several years, he or she must be a highly competent professional. Unfortunately, there are people that have been in the business for 30 and 40 years who are unable to distinguish between synthetic and natural stones or judge the quality of colored gems. When it comes to appraising jewelry, the type and quality of ones experience is far more important than the length of it.

Appraisal Fees

As a consumer, you have the right to know in advance the approximate cost of an appraisal. Occasionally, an appraiser will tell a caller that it is unethical or unprofessional to quote prices over the phone. This is not true. Professional appraisers should at least be able to tell you their hourly fee and/or their minimum charge if they have one. Some will tell you a flat or approximate appraisal charge for the piece when you describe it to them over the phone.

There are some people that will offer to appraise your jewelry free of charge, even if you haven't bought it from them. This is a sign that either they want to buy the jewelry from you or else they want to lure you into their store to sell you some of their merchandise. Professionals charge for their services, whether they be lawyers, doctors, accountants, or appraisers.

Some people who have no credentials whatsoever charge two and three times more for their appraisals than some highly qualified appraisers. There are customers who mistakenly believe that anybody who charges a high price must be good. It's important for you to check an appraiser's qualifications.

Appraisal fees are charged in a variety of ways. Some are listed below.

- A flat fee per item, sometimes a lower fee for each additional piece brought in at the same time
- An hourly rate (often combined with a minimum fee)
- A rate fully or partly based on the carat weight of the stone(s)
- A rate fully or partly based on the gem type of the stone(s)
- A rate based on the types of reports you wish to have included in the appraisal
- A percentage rate of the appraised value of your jewelry. The higher the value, the more money the appraiser earns. If you want an appraisal that is as objective as possible, avoid appraisers with this type of fee structure. Percentage fees are seldom encountered in the United States now. The Internal Revenue Service doesn't recognize appraisals done by people who charge this kind of fee, and appraisal organizations have condemned it.

What Does the Appraisal Include?

Basically this question asks for a description of the appraisal you'll receive. The answer will tell you something about the quality of the appraiser's work, and it will help you to better compare appraisal fees. It's understandable that a nine-page report with a photo would cost more than one with only a two-sentence description and an appraised value. Some of the things appraisers might say they include with their reports are:

- The identity of the stones
- The measurements and estimated weights of the stones. (If you can tell appraisers the exact weight of the stones, this will help them give you a more accurate appraisal. Therefore, when buying jewelry, ask stores to write on the receipt any stone weights listed on the sales tags).
- An indication of any treatments used on the stones when evidence is present
- A description of the clarity, color, shape, cutting style, proportions, and finish of the stones
- Plots of the inclusions in the stones (of either all or only the major stones)
- A color analysis of the colored stones
- A country of origin report

- A metals test
- Approximate weight and description of the mounting
- A cleaning and inspection of the piece
- A photograph
- A list of the tests performed and the instruments used.

Suppose you ask someone what their appraisal fee includes and they start specifying a few of the items above. Then suppose you ask someone else and they reply, "The value and a description of the piece. What more do you expect?" Haven't those two answers helped you determine who is probably most qualified to appraise your jewelry?

It's not necessary for an appraisal to include all of the above items, and rarely is the country of origin of stones given. However, if appraisers say they include the origin when possible, this is a sign they might have specialized knowledge of colored stones such as rubies, sapphires, and emeralds.

Besides knowing what appraisers' fees include, you should know what their appraisals look like. Have them show you a sample when you go to their office, and check it for thoroughness and professionalism. If you are getting an insurance appraisal, you'll want it to include as much specific information as possible. This will increase your chances of having lost or stolen jewelry replaced with items of the same type and quality.

Appraising Jewelry While you Wait

Not all appraisers have a policy of appraising jewelry while you wait. However, it may be inconvenient for you to make two separate trips for an appraisal--one to drop off the jewelry and one to pick it up. Most appraisers will try to accommodate your needs. Some appraisers will only appraise it in front of you. However, they often send you the final written appraisal afterwards. No matter what their policy, you usually need to make an appointment.

Even if it doesn't matter to you whether you leave the jewelry with them or not, it's not a bad idea to ask appraisers if they do on-the-spot appraisals. Their answers will give you added information about them. For example, look at the three responses below. These are actual answers to the question, "Can you do the appraisal on the spot?"

- "Only if you pay double because it will keep us off the floor away from selling."
- "Yes, but my schedule is limited. What I mostly do is custom design, not appraising."
- "Yes, but it will have to be in the afternoon. I reserve mornings for jewelers and their questions and appraisals."

Which of the three is probably most qualified to appraise a $9000 sapphire ring? Which of the three might be most qualified to appraise custom designed gold jewelry?

Independent Appraisers versus Jewelry Store Appraisers

Most jewelry appraisers are involved in some way in the buying and selling of gems. This is valuable experience that gives them first-hand information about market prices. Appraising becomes an additional service they can offer their customers. If you need an appraisal for insurance, estate, or tax purposes, the jewelry store where you normally shop may be the ideal place to have your jewelry appraised. However, before you have a store appraise jewelry with center stones of ruby or sapphire, ask to see a sample appraisal of a piece set with a colored stone (a gemstone which is not a diamond). Check it for thoroughness. Does it include an analysis of the color, clarity, proportioning, and brilliancy of the stone(s) as well as its weight, measurements, and a description of the mounting? As mentioned earlier, an insurance appraisal should contain as much pertinent information as possible. This will help you get a replacement piece of the same quality instead of one of lesser value.

There are times when it's best to go to an independent appraiser (someone who does not buy or sell jewelry or gems). An example would be when you want a second opinion, as in the case of Joe with his $9000 sapphire ring.

If Joe goes to another jewelry store, he will put them in an awkward position. If their appraisal value of the ring is higher than what Joe paid, they may make their merchandise appear over-priced. If it is lower, they may make another jeweler look bad. (The chances of the appraised value being the same as the purchase price are almost nil due to variations in gold and colored-stone prices and differences in mark-ups and labor costs.) Some stores will intentionally give a low appraisal in order to turn around and sell their own merchandise. Therefore, the best solution in a case like this is to go to someone who has nothing to gain or lose from the appraised value he gives.

An independent appraiser gets pricing information from dealers, jewelers, manufacturers, designers, auction houses, trade shows, price-guides, catalogues, magazines, and computer networks. He or she may be listed as a laboratory, gemologist, or simply as an appraisal service. Since independent appraisers often do full-time appraisal and identification work, they tend to be more skilled at identifying gems than someone who concentrates on selling. At times, even jewelry stores may rely on them-- when appraising a high-priced ruby or sapphire, for example. This requires specialized knowledge. The next section outlines why.

Why Rubies and Sapphires are Harder to Appraise than Diamonds

There may be several people in your area that could do a high-quality appraisal of a $50,000 diamond solitaire ring. There may be only one or two that could do one of a $9000 ruby or sapphire ring. Rubies and sapphires are much harder to grade, price, and identify than diamonds. Some reasons for this are as follows:

♦ Diamonds can be quickly identified with a device that measures how they conduct heat. Plus they have internal and surface characteristics which are very different from other gems.

There is no device that quickly identifies rubies and sapphires, and their blemishes and inclusions often resemble those of other stones.

- Distinguishing natural rubies and sapphires from synthetics is a constant challenge, and it is becoming more and more difficult due to new, sophisticated methods of growing them. Synthetic diamonds, on the other hand, are hardly ever encountered in jewelry. They are relatively new and at present, are produced for industrial use.

- Rubies and sapphires are almost always treated to improve their appearance, and this can sometimes affect their value. It takes a great deal of expertise to detect these treatments. Diamonds are seldom treated, although that is changing.

- Country of origin is sometimes a significant factor in determining the value of rubies and sapphires, but for diamonds, it's unimportant. Determining the country of origin is a highly specialized skill that few people have.

- It's harder to estimate the weight of rubies and sapphires because their proportions are more irregular than those of diamonds. Most corundum is cut for weight retention rather than brilliance.

- Information on grading rubies and a sapphires is not readily available. Diamond grading information, however, is easy to find.

- It's debatable as to what are the most desirable colors for rubies and sapphires. There is international agreement on how diamond color should be graded (except for fancy-color diamonds).

- There's no standardized system for appraisers to use to describe ruby and sapphire colors. For diamonds, there is.

- Ruby and sapphire pricing guides are not as widely used as those for diamonds. Therefore, they are not as reliable.

- The bulk of fine jewelry sales are diamonds and gold. Consequently, appraisers don't get as much practice identifying and valuing rubies and sapphires as they do with diamonds.

Appraising a ruby or sapphire is not easy. If a jeweler tells you, he doesn't feel qualified to appraise your ruby or sapphire and recommends someone else, this is not a sign of incompetence. It's a sign of honesty and the ability to recognize one's limitations. It's the same as a general medical doctor referring you to a specialist.

Jewelry appraising is an art. There is a lot more to it than simply placing a dollar value on a stone or jewelry piece. If your jewelry has a great deal of monetary or sentimental value, it's important that you have a detailed, accurate appraisal of it. Take as much care in selecting your appraiser as you did with your jewelry.

16

Gem Lab Documents

As Cindy was opening her mail, she pulled out an impressive-looking document from one of the envelopes. It read:

ABCD GEM LABORATORIES, INC.

Certificate of Authenticity

Description:	Genuine lab-grown ruby
Weight:	One carat
Shape:	Round brilliant
Cut:	58 Facets
Color:	Exceptional
Clarity:	Internally flawless
Finish:	Excellent

She had a look at the letter which came along with it. It said that for just $29, this beautiful stone could be hers. Plus, she could get 40% off any of the mountings shown in the enclosed leaflet. There were even testimonials from movie stars who had bought jewelry from the firm.

Cindy had always dreamed of having a ruby pendant, but she had never been able to afford one. There was no way she could pass up an offer like this, especially since it came with a certificate. She sent off for the stone and also ordered one of the 14-karat-gold pendant mountings.

This story has a happy ending. Cindy was delighted with her pendant when it arrived, and her friends raved about how attractive it looked. The mail-order firm was even more delighted. It made about a 2800% profit off of an inexpensive type of synthetic ruby (the stone had cost them less than a dollar). Plus it made a nice profit off of the mounting.

In the United States, this type of mail-order advertising is not uncommon, especially with cubic zirconia stones. Certificates are often included because they make the stone seem more valuable. However, any stone can come with a certificate no matter what its quality or price. Unlike appraisals, certificates and lab reports do not indicate the monetary value of a stone. They give details about the stone's identity, weight, shape, measurements, and sometimes its quality.

When used properly, gem lab reports can be a big help to buyers. They serve as a documented second opinion by impartial experts (when issued by reputable labs). Unfortunately, they are sometimes misused. Some of the ways they're abused are indicated below:

- A synthetic ruby or sapphire may be cut to match a stone on a report. Then it is substituted for the natural stone. Ripoffs like these can be avoided by dealing with reliable jewelers and by examining stones carefully before you buy them. See Chapter 10 for guidelines on spotting synthetic rubies and sapphires.

- An identification report from a respected lab may be used to make a very low quality stone seem valuable. If a stone is identified as a natural ruby on a report, this does not mean it is worth a lot. You need to see a quality analysis of the stone before you can make a judgement.

- Grading reports from respected labs may be used to make small, top-quality stones seem as valuable as large stones with the same color and clarity grades. The carat weight of a stone plays a major role in determining its value, so it should not be ignored.

- Salespeople may focus on just one quality characteristic on the report, such as a high clarity grade. The color, proportions, and brilliance are also important factors in determining a stone's value. Pay attention to them too.

- Occasionally the grades on a document may be altered. Most labs make it very difficult to change or counterfeit their documents. Consequently, this is seldom a problem. If you have a question about a report, you can verify the information on it by calling the lab that issued it.

- Quality-analysis reports from non-existent labs may be used to mislead buyers. The grades on these reports are often inflated. Before relying on information from a lab report, check out the lab. Interview them on the phone using the guidelines in the chapter on choosing an appraiser. Also check if it is one of the three labs listed in the next section.

Internationally Recognized Labs

If you plan to sell an expensive gem on the international market or through an auction house such as Christie's or Sotheby's, it should be accompanied by a report from an internationally recognized lab. Also, when you buy stones that are very expensive, it is a good idea to have them identified and graded by one of these labs.

There are a variety of labs known world-wide, but three that are particularly respected for both their reports on colored stones and diamonds are:

- **GIA (Gemological Institute of America) Gem Trade Laboratory, Inc.** -- Santa Monica, California and New York City
- **Gubelin Gemmological Laboratory** -- Lucerne, Switzerland
- **AGL (American Gemological Laboratories)** -- New York City

These three have gone beyond merely issuing lab documents. They have made significant contributions both to the jewelry business and to the study of gemology. A brief account of each is given below:

The **GIA** is responsible for standardizing diamond grading. It is their color and clarity grading system which is used by diamond dealers and jewelers around the world. The GIA has also been in the forefront with their research on synthetics, gem treatments, and testing procedures. The results of their findings are regularly reported in their journal *Gems & Gemology*.

The GIA Gem Trade Laboratory, Inc. offers pearl identification reports and colored stone reports which not only identify stones but also state if they have been enhanced in any way (fig 16.1). In the future, the GIA plans to add a quality analysis to these reports which will incorporate the colored stone grading system they are currently teaching in their gemology courses.

The **Gubelin Gemmological Laboratory** is renowned for its extensive research and publications on gemstone inclusions and testing methods. Its findings have led to the use of new techniques for identifying gems such as Raman laser analysis, x-ray fluorescence analysis, and ultraviolet/visible/infrared spectrometry. It is also noted for being the first to indicate on its lab documents where a gem was mined.

Besides issuing colored stone identification reports with country-of-origin statements, when possible (fig 16.2), the Gubelin Laboratory identifies diamonds and pearls. Approximately one-third of the world's largest cut diamonds have been tested by Gubelin experts.

GIA GEM TRADE LABORATORY, INC.

A Wholly Owned Subsidiary of the Gemological Institute of America, Inc.

580 Fifth Avenue New York, New York 10036 (212) 221-5858	550 South Hill Street Los Angeles, California 90013 (213) 629-5435	1630 Stewart Street Santa Monica, California 90404 (213) 828-3148

GEM IDENTIFICATION REPORT 0123456

TO Whom It May Concern:

Date March 7, 1989

FOR Identification:

DESCRIPTION

INSTRUMENTATION (When Applicable)

1 transparent green emerald cut stone measuring approx. 13.04 x 10.47 x 6.79 mm., wt. 7.91 cts.

Binocular microscope
Refractometer
Spectroscope
Polariscope
Dichroscope
Heavy liquids
Ultra-violet
Color filter
X-Radiography
X-Ray Fluorescence
Ancillary equipment and tests as necessary

SPECIMEN

CONCLUSION: ----------------------TOURMALINE, WT. 7.91 CTS.--------------------

GIA GEM TRADE LABORATORY, INC.
GIA Gem Trade Laboratory, Inc.

This report is not a guarantee, valuation or appraisal. The recipient of the report may wish to consult a jeweler or gemologist about information contained herein.

NOTICE: IMPORTANT LIMITATIONS ON REVERSE

Fig. 16.1 A sample colored stone identification report from the GIA Gem Trade Laboratory, Inc.

EDELSTEIN-BERICHT / RAPPORT DE PIERRE PRÉCIEUSE
PRECIOUS STONE REPORT

Luzern/Lucerne 31.12.1990 AP/kl

Gegenstand/Objet/Item	One unmounted gemstone
Gewicht/Poids/Weight	3,88 ct
Nr/No	9012999
Schliffart/Taille/Style	faceted top: modified brilliant cut base: step cut
Form/Forme/Shape	cushion
Abmessungen/Dimensions/Measurements mm	9,22 x 7,69 x H 6,37
Farbe/Couleur/Color	RED
Eigenschaften/Propriétés/Properties	absorption spectrum of ruby
Brechzahl/Indices de réfraction/Refractive indices	o 1,770 e 1,762
Doppelbrechung/Biréfringence/Birefringence	0,008
Pleochroismus/Pléochroisme/Pleochroism	strong, purple red, orangy red
Fluoreszenz/Fluorescence/Fluorescence	UV 365 : strong, red UV 254 : medium, red
Dichte/Densité/Density	appr. 3,99
Mikroskopischer Befund/Examen microscopique/Microscopic examination	These properties and the internal features correspond to those of a natural ruby.

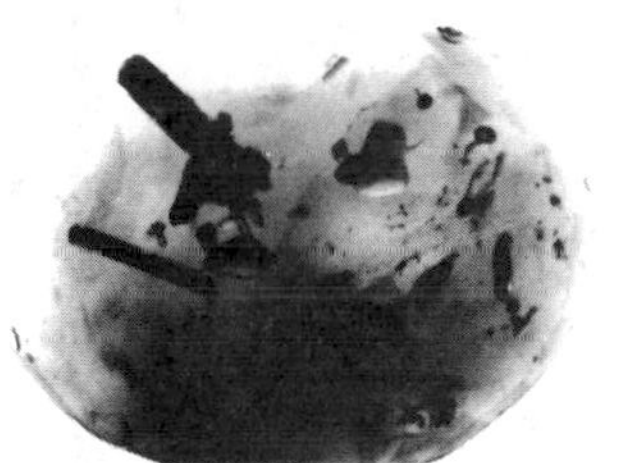

Origin : Burma

* * *

Inclusions of pargasite and spinel (see microphoto).

SPEZIES/ESPÈCE/SPECIES

NATURAL CORUNDUM

VARIETÄT/VARIÉTÉ/VARIETY

RUBY

GEMMOLOGISCHES LABOR GÜBELIN
LABORATOIRE GEMMOLOGIQUE GUBELIN/GEMMOLOGICAL LABORATORY GUBELIN
Gübelin AG, Denkmalstrasse 2, CH-6006 Luzern

A. Peretti
Dr.sc.nat.ETH, F.G.G.

S. Gübelin
Graduate Gemologist

Fig. 16.2 Note: The actual photograph is in color. In 1991, the format of the Gubelin identification report will be modified.

American Gemological Laboratories

Olympic Tower
645 Fifth Avenue
New York, N.Y. 10022
(212) 935-0060
(212) 935-0071

Colored Stone Certificate

The information contained in this report represents the opinion of American Gemological Laboratories regarding the characteristics of the colored stone(s) submitted for examination.

The conclusions expressed are the American Gemological Laboratories' interpretation of the results obtained from gemological instruments and grading techniques designed for this purpose. Conclusions may vary due to the subjective nature of colored stone analysis. Mounted stones are graded only to the extent that mounting permits examination.

CERTIFICATE NO: CS SAMPLE

DATE: 15 December 1990

Girdle Diameter = 100%
Table %
Crown
Depth %
Pavilion Depth
Girdle Thickness
Culet

Proportions: Cutting

IDENTIFICATION: Natural Sapphire, Kashmir*

SHAPE AND CUT: Cushion Antique Mixed Cut

CARAT WEIGHT: 5.96 Cts.

MEASUREMENTS: 10.26 x 9.65 x 6.74 mm.

COLOR GRADE: 3.5/80*

Color Rating/Tone 3-4/80

Color Scan $B_{70}V_{15}G_{15}$

Light Source Duro-test: Vita Lite

CLARITY GRADE: LI_2*

PROPORTIONS: Very Good-Good (3-4)

Depth % 69.8%

Brilliancy % Range: 70-100%
Average: 80%

Finish Very Good (3)

NOTE: All color determinations are subject to the color temperature of the light source and the color sensitivity of the observer.

The quality and commercial desirability of any color is determined by examining the interrelationships of all factors indicated under Color Grade. Conclusions may vary due to the subjective nature of colored stone analysis.

COMMENTS: *Total quality integration rating: Excellent. *Based on available gemological information, the origin of this material would be classified as Kashmir. *No gemological evidence of heat induced appearance modification present. *Moderate texture present.

Color Rating (AGL)

1	2	3	4	5	6	7	8	9	10
Excellent		Very Good		Good		Fair		Poor	

Tone (AGL) (0 = Colorless 100 = Black)

0	5	10	15	20	25	30	35	40	45	50	55	60	65	70	75	80	85	90	95	100
V. Light				Light			Light-Med.			Medium			Medium-Dark					Dark-V. Dark		

Clarity Scale (AGL)

FI	LI_1	LI_2	MI_1	MI_2	HI_1	HI_2	EI_1	EI_2	EI_3
Free of Incl.	Lightly Included		Moderately Included		Highly Included		Excessively Included		

Proportions/Finish (AGL)

1	2	3	4	5	6	7	8	9	10
Excellent	Very Good		Good			Fair		Poor	

NOTICE TO CLIENTS: The client agrees to be responsible for and to protect, indemnify and hold harmless American Gemological Laboratories, Inc., and its officers, directors, and employees from and against any and all injury, loss, damage, destruction or otherwise of said gem materials while in the possession of American Gemological Laboratories resulting from any cause not attributable to the negligence of American Gemological Laboratories or its employees.

The client agrees that this report is for his/her exclusive use.

Neither the American Gemological Laboratories nor any of its employees shall be responsible for any action that may be taken on the basis of this report.

This Certificate or the name of American Gemological Laboratories, Inc. may not be reproduced in whole or in part without the express, prior written authorization from American Gemological Laboratories.

Do not accept copies, alterations or corrections on this document for final transactions.

This report prepared by: **American Gemological Laboratories, Inc.**

C. R. Beesley, Jr.

C. R. Beesley, President

Fig. 16.3 A sample colored stone certificate from the American Gemological Laboratories

AGL was the first lab to grade colored stones for quality and to routinely note gem enhancements on their identification reports. Included on their colored stone certificates is the stone's identity, country of origin (when possible), and a thorough analysis of the stone's clarity, proportions, and color (fig 16.3). The color description is based on the Color/Scan grading system they developed.

AGL has been an outspoken advocate of the consumers' right to full information about their gem purchases, and it has been particularly involved in pushing for disclosure of gem treatments.

When these labs created their documents, they intended for them to be an aid to buyers and a deterrent to fraud. Unfortunately, their lab reports have sometimes been used to defraud the public. In order to gain maximum benefit from these reports and at the same time avoid their pitfalls, keep in mind the following suggestions:

- **Do not buy expensive jewelry and gems through the mail or over the phone if you don't know the seller.** These are probably the most common situations in which lab documents are misused.

- **Avoid gem investment schemes even when the stones come with lab reports.** People have lost their life savings by believing promises of high returns on gem investments.

- **Keep in mind that lab documents are not infallible.** They only represent the opinions of the labs issuing them. To emphasize this, the GIA and Gubelin lab specify that their grading documents be called reports, not certificates. Jewelers and dealers, however, often refer to them as "certs" or "certificates" because these are the terms they are accustomed to using. No matter what reports may be called, reputable gem laboratories never guarantee them.

- **Do not buy gems solely on the basis of a lab report.** Always examine the stones yourself with and without magnification before you buy them. Occasionally, the stone may be different than the one on the report, or a stone might have been damaged since the report was issued. Even more important, though, is the fact that a written report cannot give a complete picture of a gem. You have to see the stone to really know what it looks like. Lab reports were never created to be a substitute for viewing a stone. Use them as an aid to judging quality and as a confirmation that a stone is real and natural. But when it's time to make the final choice, you be the judge.

17

Tips for Travelers

The art of imitating gems has reached a high degree of perfection, and while the substitutes thus prepared have legitimate uses, the temptation to palm them off on the unsuspecting for real gems, at or near the price of the genuine, is often too strong to be resisted. . . Tourists are especially liable to deception of this sort, since their purchases must be largely made of itinerant venders, with whom they are not acquainted. The Persian turquoise venders, knowing the liability of some of their wares to fade, are accustomed to leave for parts unknown as soon as their stock is disposed of, and gem-sellers of other nations often exhibit similar propensities.

(1903, Dr. Oliver Farrington, curator at the Field Museum in Chicago, from his book *Gems & Gem Minerals*)

Nowadays, tourists are more likely to buy gems at shops than from itinerant vendors. Nevertheless, the risk of deception is still present. Tourists who expect to find gems at a mere fraction of their cost back home are especially vulnerable. For example, they may jump at the chance to buy a "ruby" for 1/10 of what it would normally cost. They should instead be suspicious. Natural rubies are an international commodity, and there is a limited supply of them. It would make no sense for a store to sell a ruby at such a discount.

It's true that gems often sell for less in countries where they are mined, especially when labor is cheap there. Good buys, however, can also be found outside of these countries. For example, around 1987 after the dollar was devalued, Thai dealers often found better bargains on rubies and sapphires in the United States than in Bangkok. Even now, due to intense competition, large volume buying, and a variety of other factors, prices on ruby and sapphire jewelry in North America may compare favorably with prices in Asia. There are other reasons besides price to consider when deciding whether to buy abroad or at home. These are listed on the following page.

Advantages of Buying Ruby & Sapphire Jewelry Abroad

- The selection may be better, especially with respect to fancy color sapphires, star stones, or high quality rubies and blue sapphires. Most jewelry stores in the US, for example, don't stock as much ruby and sapphire jewelry as stores in Thailand. In the US, there is a greater demand for diamond jewelry. Consequently, stores there tend to invest more in diamonds than in corundum.

  Jewelry abroad may also come in distinctive styles that are hard to find elsewhere. Buying these styles ready made is not only cheaper but also more convenient than having them custom made in your own country. Sometimes, however, the styles are designed more to suit local tastes than those of foreigners. Don't buy a style just because it's different. Make sure you like it.

- Fine jewelry makes an ideal souvenir. It's relatively small and usually lightweight, so you don't have to worry about excess baggage when you buy it. It represents a product that's either mined or crafted in the country you are visiting. Also, unlike many other souvenirs, it's not apt to be viewed as junk when you get it back home or give it as a gift.

- It may be more convenient and enjoyable to buy it on a trip than to shop for it at home. People often have more time to shop when they're on vacation, and they tend to be in a more relaxed state of mind. Instead of being a chore, shopping becomes a form of recreation. Finding places to shop may be easier, too. In Asia, lots of shops may be concentrated in one area, making it convenient to find what you want.

Advantages of Buying Jewelry at Home

- The service is often better. At home, jewelers not only look after you during the sale, many will continue to service you afterwards. They may periodically check the stones and clean the piece free of charge. They may also be available to update appraisals and answer your jewelry-related questions. Some will even allow you to trade in the piece for its full retail value.

  Because they offer continuing service such as free repairs for defective jewelry, jewelers at home have more incentive to sell pieces that are sturdy and problem-free than stores which concentrate on one-time sales to foreigners.

- If problems arise, it's easier to resolve them at home. When jewelry is misrepresented or defective, it's a bigger hassle to return it to a jeweler abroad than to one at home. Also, there may not be as much consumer protection outside of one's country. It's much easier to take legal action against someone nearby than thousands of miles away.

- It may be more convenient to buy jewelry at home. Perhaps you have a hectic travel schedule with little time for shopping. Rushed purchases are often regretted later on. At home you may have more time and a wider range of reliable jewelers to choose from.

Maybe you don't have any travel plans at all, but you're interested in a ruby or sapphire. There is no need to fly abroad to make the purchase. There are probably plenty of good buys right at home.

Tips on Buying Gems Abroad

Before leaving home

- **Learn as much as possible about the gems you're interested in.** Knowledge is your best defense against ripoffs.

- **Shop around to see what's available locally.** Look at a wide range of qualities, and check the prices. If you limit yourself to looking at commercial quality goods at a local discount store, you might be misled into thinking that the better quality goods abroad are overpriced when, in fact, they might be lower in price.

- **Store your valuable jewelry in a secure place** such as a safe deposit box. If you travel with it, you risk losing it and being questioned by customs about it on your return home.

- **If you do travel with jewelry from home, take along a photocopy of the purchase receipt or appraisal.** It's up to you upon returning to prove to customs that you did not buy it abroad. Unlike articles such as cameras, it may not be possible to register jewelry with customs. If you don't have receipts or appraisal forms, photograph the jewelry piece next to something found only in your country, such as your car license plate.

Choosing a jewelry store

- **Is the store recommended by a knowledgeable and trustworthy person?** Buying jewelry abroad is risky. Therefore, it's a good idea to get recommendations. Also local chambers of commerce and jewelry associations may be able to give you information about jewelry stores.

- **Does the staff seem knowledgeable, and does someone in the store have gemological credentials** such as those listed in the chapter on appraising? It's important, for example, that someone on the staff be able to detect synthetic stones. Even honest jewelers may end up selling synthetics as natural stones if they are unable to identify them.

- **Do the salespeople tell you both the good and weak points of their stones?** Ethical professionals want you to understand what you are buying; so they will show you how to judge quality, and they will give honest evaluations of their merchandise.

- **Does the store offer a 100% money-back guarantee?** Some jewelers will only allow exchanges in order to prevent customers from "borrowing" their merchandise for a special event and then returning it afterwards for a full refund. When you buy jewelry away from home, though, exchanges become impractical. There are many jewelers abroad that offer and honor 100% money-back guarantees. It's best to deal with one of them. That way, if a competent appraiser at home recommends returning your purchase, it will be easier to get a refund.

♦ **Does the staff treat you with patience and courtesy?** Their attitude is just as important as their technical knowledge.

Choosing the stone(s)

♦ **Remember to view the stone(s) under different types of light**--fluorescent, incandescent (light bulbs), and daylight near a window. The color of the gems can look different under the various lights.

♦ **Look at the stone under magnification.**

♦ **Pay close attention to the quality of the cut.** The colored stones sold in certain countries tend to have exceptionally large "windows." (See Chapter 8 for tips on judging cut.)

♦ **If possible, avoid closed-back settings** (These are discussed in the chapter on imitations). They are frequently an indication that something is being hidden.

♦ Follow the guidelines listed in the chapters on judging color, cut, and clarity.

Making the purchase

♦ If you have any doubt whatsoever about the purchase, pay with a credit card. Resist a bargain for cash. Credit card companies can be an excellent source of protection. When you get home, have a competent appraiser or lab check the stone or jewelry piece. If there's a problem that the store refuses to correct, notify the credit card company, explain the situation, and ask them to stop payment. Before your trip, however, you should ask your credit card company what their policy is regarding fraudulent merchants. Some companies may expect you to pay even when a store defrauds you or breaks its agreements. Others may offer protection only if the purchase was made within a specified distance from your home.

Consider, too, having the piece checked by a qualified lab in the city of purchase. It's easier to get your money back on the spot than thousands of miles away. Be wary if a store is overly pushy about using a specific appraiser. They may be working together.

♦ **Know how much the purchase is in your home currency**. Foreign currency prices can be confusing.

♦ **Get a detailed receipt and have verbal agreements put in writing** (for example, have them write out "100% money-back guarantee"). The receipt should at least include the identity and shape of the stone(s), the identity and purity of the metal, the carat weight of any major stones and the total weight of any smaller stones of each gem type. For added protection, have the store specify **natural** ruby or **natural** sapphire on the receipt. Technically synthetic ruby, for example, is ruby. Ethically, though, it should be called synthetic or lab-grown ruby. (In the US, it would be against the law to call a synthetic ruby simply a ruby. The gem name must be qualified with terms such as "synthetic" or "lab-grown." Not all countries, however, have laws like this).

If you're buying a loose stone, ask the store to include its millimeter dimensions. This information will help identify it later on. Incidentally, receipts in Asia often identify light red corundum stones as rubies. On appraisals and lab reports in Europe and North America, the same stones will probably be called pink sapphires.

- **Get the store's phone number, street address, and fax number** (if they have one). You may need to contact them later.

- **Ask the store if it has an outlet in your home country that can service you and make refunds.** If it does, get the name, address, and phone number.

After the purchase

- **Keep the jewelry or stone in a secure place such as a money belt or a hotel safe deposit box.** Be sure, too, it's in a protective pouch or covering so it doesn't get scratched. When flying from one country to another, it might be better to wear the jewelry than to put it in your carry-on-bag or purse. (In certain places, however, such as some countries in South America, you should avoid wearing necklaces and bracelets. They are apt to be snatched off of you). Do not pack valuable jewelry in your checked baggage. It can either be lost or stolen.

- **Be careful at airport security check points**, especially when valuable jewelry is in your carry-on-bag or purse. While you are being body searched, someone may either steal things out of your carry-on-bag or else steal the whole bag. Either have someone keep an eye on your bags or wear the jewelry as you are searched.

- **If you bought jewelry at a store a tour director recommended, be sure to get his or her address and phone number.** Tour directors can be a big help if problems arise later on. (A tour director is the person who accompanies you throughout a tour and represents the company you booked the tour with. Local guides, on the other hand, give you commentary about their city or country. Unlike the tour director, guides are not necessarily held responsible by their employers for the reliability of the stores they recommend.)

- **Keep all receipts handy and make a record of your purchases for customs.** Sometimes people are overly concerned about customs duties. On loose stones there may be little or no duty, and the duty on jewelry may be less than what the sales tax on it would be back home. Laws regarding customs duties change frequently. Check with customs in your home country for current detailed information regarding duties. Direct information from customs is more reliable than what you'll hear from jewelers abroad.

Incidentally, jewelry bought in duty free shops must be declared, and customs will charge you the same duty on it as on other jewelry from the same country. "Duty free" normally means that no import or sales tax has been included in the price. Jewelry sold in duty-free shops is not necessarily cheaper than elsewhere. In fact, often the opposite is true.

When Problems Arise

If an appraiser or gem lab tells you your purchase abroad is either overpriced or misrepresented, proceed as follows:

- First send the store a photocopy of the appraisal or lab report (by fax or air mail), and explain how you would like the matter resolved. It's a good idea to follow up with a phone call. You should let the store have a chance to give you an explanation. Appraisers are not infallible, nor are they all highly qualified. In fact, some have little experience valuing and identifying corundum.

- If the store ignores you or refuses to resolve the matter, tell the owner or manager that you will be filing formal complaints with business and trade organizations in his city and with your credit card company (if you paid by credit card). If the store still refuses to cooperate, then follow through on your threats.

- To get the addresses of business and jewelry organizations, try calling the trade commissioner that represents the country where you made the purchase. Explain the problem, and ask if he can help you. You can usually get the phone number of the trade commissioner by calling the local consulate.

- If you paid by credit card, phone the credit company within 30 days if possible, explain the problem, and ask them to stop payment on your purchase. They'll tell you if they need documentation and how to return the merchandise. Get the name of the person on the phone and the file or reference number of the case (if one can be assigned).

- If you bought the jewelry from a store a tour director recommended, you may want to contact him or her first. Good tour directors can save you a lot of time and hassle if problems arise. They will contact the jeweler for you, explain the problem, and relay your requests. Either the jeweler or tour director or both will get back to you to take care of the matter.

- If you bought the jewelry from a store recommended in a brochure of a reliable tour operator in your country, contact the tour company. They'll give the jeweler one of two choices--either resolve the matter or lose all future business from them.

- When you return merchandise, normally, it's best to send it insured airmail and to pay for a return receipt. You may need proof that you returned the merchandise.

- Keep originals of all documentation regarding your case. Only send copies.

Tips on Having Jewelry Custom Made

If you don't find the style of jewelry you are looking for, you may wish to have it custom made. Many Asian countries are noted for their fast service--often 1 to 2 days. They can be this quick because they have more jewelers per capita than countries such as the US. A jeweler in Asia may have only one or two pieces to work on at a time, whereas one in the US may be handling 30 or 40 jobs at once. When having jewelry custom made, follow the guidelines below:

- Try on jewelry pieces that resemble the one you want to have made. What looks good in a picture may not look good on you.

- If possible, have good drawings, photos, or models of the jewelry piece you want made. Never assume that the jeweler understands your verbal description of what you want. Be as specific as possible about how you want the jewelry to look.

- Don't assume a piece of jewelry will look exactly as it does in a photo. It should however, have a close resemblance.

- Always tell the store you need the jewelry piece earlier than you actually do, especially if it's a complicated job. In Asia, you should allow 3 or 4 hours extra; and in countries such as the US, at least 3 or 4 days. Either the jewelers may not finish on time, or alterations may be needed.

- Get a written estimate of the cost of the jewelry. If more stones are needed than estimated, the jeweler is not expected to give them to you free of charge. He should, however, get your permission before doing anything that would increase the estimated cost of the jewelry.

- Know the refund policy of the store. It is normal for a store to retain at least a portion of your deposit if you decide not to buy the jewelry you ordered, particularly if it's a style that would be difficult to sell. Pay deposits with credit cards. If the store does not deliver the job as promised, it will be easier to get your money back.

When you have a piece custom made, it often means more to you than ready-made jewelry. The piece is unique, and you played a role in creating it. The experience of having jewelry made should be a positive one. Prevent it from turning into a negative one by taking the necessary precautions.

18

Caring for Your Ruby & Sapphire Jewelry

Cleaning the Stones

It's not a coincidence that the ruby and the sapphire are the most popular colored stones for wedding rings. Their hardness, toughness, and resistance to chemicals make them ideal for everyday wear. Unlike emeralds, pearls, opals, turquoise, tanzanite, and some other gems, rubies and sapphires are not damaged by steam cleaners and ultrasonics (cleaning machines that shake dirt loose with a vibrating detergent solution using high-frequency sound waves).

If your ruby or sapphire jewelry hasn't been cleaned for a long time and is caked with dirt, it would be best to have it professionally cleaned by one of these methods instead of trying to clean it yourself. There are a few cases, however, where you should avoid having corundum cleaned in an ultrasonic or steam cleaner, namely with:

- Badly flawed stones--they can be further damaged.
- Black star sapphires--some tend to be fragile and may split.
- Rubies with glass-filled cavities--the filling may fall out.
- Oiled and/or dyed stones--The oil and dye may be removed. (Highly flawed corundum is occasionally oiled and dyed.)

Risky cleaning procedures can be avoided if you clean your jewelry on a regular basis. Simply soak and wash it in warm sudsy water using a mild liquid detergent. Then dry it with a soft, lint-free cloth. If the dirt on the rubies or sapphires cannot be washed off with the cloth after soaking, try using a toothpick, a Water Pik, or unwaxed dental floss to remove it.

No matter how dirty your stones might be, never boil them because this may create cracks or a change in color. Also never soak them in any solution containing borax. This can cause the surface of the stones to corrode.

Cleaning Gold Mountings

Gold is over 800 times softer than rubies and sapphires, so it can be scratched more easily. If possible, avoid using brushes on gold when you wash it in soapy water. Rub the metal instead with a soft cloth. Also avoid cleaning it with toothpaste or powder cleansers because these can wear away the metal.

Since jewelry gold is not pure, a variety of chemical products may discolor or dissolve it. A few of these products and their affect on gold alloys are listed below:

- **Chlorine**--it can pit and dissolve the metal, causing prongs to snap and mountings to break apart. Afterwards, it might appear as if you've been sold defective or fake gold jewelry. Therefore, avoid wearing gold jewelry in swimming pools or hot tubs that have chlorine disinfectants, and never soak it or clean it with bleach.

- **Lotions and cosmetics**--besides leaving a film on the jewelry piece, they can tarnish it, especially where it has been soldered together. If possible, put your jewelry on last, after applying make-up and spraying your hair.

- **Perm solutions**--they have a tendency to turn solder joints black. In some cases, the whole jewelry piece has taken on a brownish cast.

- **Some medications**--they may cause a chemical reaction in certain people. This can make their skin turn black when it comes into contact with the gold alloy.

- **Polishing compounds**--They can blacken your skin if they remain on the gold. Polishing cloths sold in jewelry stores may contain a mild abrasive for shining the metal. When using these cloths, be sure to wash or wipe the metal thoroughly afterwards.

Storing Your Jewelry

When you store jewelry, protection from theft and damage should be a prime consideration. A jewelry box can protect pieces from damage if they are stored individually, but it is one of the first places burglars look when they break into a home. So it's best to reserve jewelry boxes for costume jewelry when they are displayed on tables or dressers.

Jewelry pieces should be wrapped separately in soft material or placed individually in pouches or the pockets of padded jewelry bags. If a piece is placed next to or on top of other jewelry, the metal mountings or the stones can get scratched. Use your imagination to find a secure place in your house to hide jewelry pouches, bags and boxes. If jewelry is seldom worn, it's best to keep it in a safe deposit box.

Other Tips

- If possible, avoid wearing jewelry while participating in contact sports or doing housework, gardening, repairs, etc. Even though rubies and sapphires are very durable, they can be chipped, scratched, and cracked. If during rough work, you want to wear a ring for sentimental reasons or to avoid losing it, wear protective gloves. Hopefully, your ring has a smooth setting style with no high prongs.

- When you set jewelry near a sink, make sure the drains are plugged or that it's put in a protective container. Otherwise, don't take the jewelry off.

- Occasionally check your jewelry for loose stones. Shake it or tap it lightly with your forefinger while holding it next to your ear. If you hear the stones rattle or click, have a jeweler tighten the prongs.

- Avoid exposing your jewelry to sudden changes of temperature. If you wear it in a hot tub and then go in cold water with it on, the stones could crack or shatter. Also keep jewelry away from steam and hot pots and ovens in the kitchen.

- Take a photo of your jewelry (a macro lens is helpful). Just lay it all together on a table for the photo. If the jewelry is ever lost or stolen, you'll have documentation to help you remember and prove what you had.

- About every six months, have a jewelry professional check your ring for loose stones or wear on the mounting. Many jewelers will do this free of charge, and they'll be happy to answer your questions regarding the care of your jewelry.

19

Finding a Good Buy

Ed is shopping for a sapphire ring and has just finished reading *The Ruby & Sapphire Buying Guide*. While looking in a store window, he spots a stone with an unusually fine blue color. He decides to go in and have a look. Marian, the owner of the store, is impressed that Ed could pick out the best sapphire in her window display and gladly shows it to him. Unfortunately, the stone is way out of Ed's price range. However, Marian is able to help him find an attractive sapphire that he can afford and a ring mounting to set it in.

Even though Ed has been to a variety of jewelry stores, almost all of the sapphires he's seen have had a very dark navy blue color and hardly any brilliance. Marian sells these types of sapphires too, but she is the first jeweler to readily admit to him that many of her stones are not of high quality. She is also the first jeweler to show Ed how the color, clarity, and brilliance of a sapphire affects its value, using examples from her inventory.

Ed appreciates Marian's honesty and the time she has spent with him; he realizes that not only has he found a good quality sapphire, he's found a good jeweler as well.

Dee is on vacation in Hong Kong and wants to get a custom-made ruby pendant from a jeweler that her boss has recommended--Mr. Wong. Before she left on her trip she read *The Ruby & Sapphire Buying Guide*. As Mr. Wong shows her some stones, she realizes that a large, good quality ruby is beyond her means; she decides, instead, that a pendant with small stones would be a more affordable option.

Together, they work out a design for the pendant. Mr. Wong then brings out a packet of stones which seem to be similar in color. Dee notices, however, that their clarity and brilliance varies even though their per-carat price is the same. She mentions this to Mr. Wong and he helps her select the best quality stones to set in the mounting.

Two days later, Dee picks up the pendant. She's pleased with the way it looks, but she's even more pleased that she's had a role in creating it.

Shopping for rubies and sapphires turned out to be a positive experience for Ed and Dee, but this was largely because they took the time to learn about these stones beforehand. Listed below are some of the guidelines that helped Ed and Dee and that can help you.

- Know what types of flaws and cutting defects to avoid. These are outlined in Chapters 7 and 8. Without paying any extra, Dee got the best rubies in the packet because she was able to spot quality differences. Mr. Wong was motivated to help her select the best stones because he knew that she would appreciate his extra effort. The lowest quality stones in the packet will most likely be sold to people who are unable to evaluate quality.

- When judging prices, try to compare stones of the same shape, size, color, clarity, and cut quality. All of these factors affect the cost of rubies and sapphires. Due to the complexity of colored-stone pricing, it's easier for consumers to compare stones that are alike.

- Compare the per-carat prices of stones rather than their total cost.

- Before buying a ruby or sapphire, look at a wide range of qualities and types. This will give you a basis for comparison.

- Be willing to compromise. Both Ed and Dee had to settle for something less than they really wanted because their pocketbook didn't match their tastes. Ed had to get a sapphire of lower quality than he would have liked. Dee had to get several small rubies instead of a large one. Even people with unlimited budgets have to compromise sometimes on the size, shape, color, or quality due to lack of availability.

  Rubies and sapphires don't have to be perfect for you to enjoy them. However, if perfection is your goal, you might wish to consider buying a synthetic stone.

- Beware of sales or ads that seem too good to be true. The rubies or sapphires that are advertised might be of unacceptable quality, or they might be laboratory grown. They might also be stolen or misrepresented. Jewelers are in business to make money, not to lose it.

- If possible, establish a relationship with a jeweler you can trust and who looks after your interests. He can help you find buys you wouldn't find on your own.

- Place the ruby or sapphire in the groove between your fingers and look at it closely. Then answer the following questions. (A negative answer to any one of the questions suggests the stone is a poor choice.)

  a. Does most of the stone reflect light and color back to the eye? In other words, does it have "life" and brilliance?
  b. Does the color of the stone look good next to your skin?
  c. Does the stone look like a ruby or sapphire? There is no point, for example, in buying a sapphire that looks like black onyx when you can get real black onyx for much less.

Ed and Dee are not gemologists, but they have learned enough to make intelligent choices. Because they are knowledgeable, they are more likely to be treated fairly and get better service from jewelers. It's the same as if Ed and Dee were to hire a cab in a foreign city and give the driver the impression that they are familiar with his area. Chances are the ride would be more direct and the taxi fare would be less than if they acted lost.

Perhaps some readers were expecting *The Ruby & Sapphire Buying Guide* to tell them exactly what type of stone to buy. There is no one kind of ruby or sapphire that's right for all people. Choosing a gemstone is a very personal matter. *The Ruby & Sapphire Buying Guide* was written to help you make your own buying decisions, not to dictate what you should buy. It should not be used as your sole source of information. You should also talk to jewelry professionals, look at stones whenever possible, and read other literature. Above all, have faith in your intuitions and in your ability to learn to evaluate gems. When it comes to selecting a ruby or a sapphire, you're the one who knows what's best for you.

Appendix

Chemical, Physical, & Optical Characteristics of Rubies & Sapphires*

Chemical composition	Al_2O_3 (aluminum oxide)
Mohs' hardness:	9
Specific gravity:	4.00 (+.10, -.05)
Toughness:	Excellent, except for repeatedly twinned or fractured stones
Cleavage:	None
Parting:	Rhombohedral or basal
Fracture:	Conchoidal, uneven
Streak:	White or colorless
Crystal system:	Hexagonal (trigonal)
Crystal Habits:	Hexagonal bipyramid, tabular hexagonal prism
Optic Character:	Doubly refractive, uniaxial negative. Aggregate reaction common in star corundum
Refractive Index:	1.762-1.770 (+.009, -.005)
Birefringence:	.008 to .010

* Sources of data are listed at the end of the appendix.

Dispersion:	.018	
Luster:	Polished surfaces are vitreous to subadamantine. Fracture surfaces are vitreous.	
Phenomena	Asterism (6 & 12 rays), chatoyancy (very rare), color change from blue to purple or violet, green to reddish brown (very rare)	
Dichroism:	Ruby:	Purplish red & orangy red
	Sapphire:	
	Blue:	Violetish blue & greenish blue
	Yellow:	Yellow & light yellow or greenish yellow
	Orange:	Orange or brownish orange & light orange
	Green:	Green & yellow-green
	Purple:	Violet & orange
Chelsea-filter reaction:	Ruby:	Strong red
	Natural Sapphire:	
	Blue:	Blackish
	Green:	Green
	Purple:	Reddish
Absorption spectra:	Ruby:	Fluorescent doublet at 694.2 & 692.8, narrow lines at 668 & 659.2, broad band from 620 to 540, narrow lines at 476.5, 475, & 468.5nm, and general absorption of the violet.
	Natural Sapphire:	
	Blue:	Three bands at 451.5, 460, & 470nm. Kashmir and heat-treated stones often show no lines.
	Yellow:	(Australian) 450, 460, 470nm, no characteristic spectra in stones from other sources.
	Green:	450, 460, 470nm.
	Purple & Padparadscha:	May show a combination of the ruby and sapphire spectra
Cause of color:	Red:	Chromium (but sometimes iron and titanium are present and modify the color)
	Natural Sapphire:	
	Blue:	Iron and titanium
	Yellow:	Iron and/or color centers
	Orange:	Iron and/or color centers (traces of chromium in padparadscha)
	Green:	Iron or iron and titanium
	Purple:	Chromium, iron, and titanium
	Color change:	Chromium, iron, titanium, & sometimes vanadium

Ultraviolet fluorescence	Ruby:	(LW) strong to weak red or orange-red, (SW) moderate red or orange-red to inert. Varies according to place of origin.
	Natural Sapphire:	
	Blue:	(LW) Strong red or orange to inert (SW) moderate red or orange to inert depending on origin, some heat-treated or Thai stones are chalky green to SW.
	Yellow:	(LW) Moderate orange-red or orange-yellow to inert. (SW) Weak red or yellow-orange to inert.
	Orange:	Usually inert, may be strong orange-red to LW.
	Green:	Usually inert, weak red or orange in rare cases
	Pink:	(LW) strong orange-red, (SW) weak orangy-red
	Purple & Color Change:	(LW) strong red or orange-red to inert, (SW) weaker.
Reaction to heat	Infusible before a blowpipe or flame of jeweler's torch. Ruby may become green when cooling from high temperatures but turns red again when completely cooled. Heating sometimes improves the color of rubies and sapphires. It can also remove the color of sapphires permanently if they are heated to sufficiently high temperatures.	
Reaction to chemicals	Highly resistant but soldering flux or pickling solution with borax can dissolve the surface of the stone.	
Stability to light	Stable except for irradiated yellow and orange sapphires which fade	

Synthetic Corundum Types (Key Separations)

Verneuil (flame fusion)

Spectra	Sapphire:	
	Blue:	No iron lines or typical spectrum
	Green:	530 & 687nm lines
	Color change:	474 nm line
	Yellow, Orange:	No iron lines, sometimes a 690 nm line
Ultraviolet fluorescence	Ruby:	(LW & SW) Usually a stronger red than both natural and other synthetic types of rubies.
	Sapphire:	
	Blue:	(LW) Usually inert, some stones weak to strong red or orange-red,. (SW) Usually weak to moderate chalky blue or green, some stones weak to strong red, orange red or pink red.
	Orange:	(LW & SW) Inert to strong red or orange.

Chelsea-filter reaction	Sapphire:	
	Green:	Red as compared to green in natural green sapphire
Inclusions	Curved growth lines, tiny and large spherical or "stretched"gas bubbles that occur singly or in groups, whitish unmelted particles of AL_2O_3, occasionally small dark red crystals (concentrations of the coloring agent).	

Czochralski (Inamori)

Fluorescence	Ruby:	(SW) Usually much stronger red than natural
	Sapphire:	
	Orange:	(LW) strong orange-red, (SW) Weak pinkish orange.
Inclusions	Wispy whitish clouds, gas bubbles, faint curved growth lines, rain-like particles. (Czochralski corundum is usually inclusion-free.)	

Floating zone (Seiko)

Inclusions — Clouds of gas bubbles, swirled and curving growth or color zoning which may have a foggy Kashmir-like appearance. (This synthetic corundum is noted for its freedom from inclusions.)

Flux growth

Fluorescence	Ruby:	(SW) usually stronger orangy red than natural

Chatham

Spectra	Sapphire:	
	Blue:	A single diffuse band at 451.5nm, absent in some stones. (The presence of 3 iron lines indicates natural origin).
Inclusions	Platinum plates and needles, white or slightly yellowish flux of high relief often in the shape of wispy veils, seed crystals and accidental crystals of chrysoberyl, twinning and repeated twinning, irregular color swirls in ruby, faint blue to violet streamers of lines in ruby, straight angular color zoning which is particularly strong in blue sapphire.	

Kashan

Ruby color	Tends to resemble that of Thai rubies or red spinel. The orangy-red pleochroic color found in natural rubies is often more yellowish in Kashan ruby.
Inclusions	Flux fingerprints, feathers, and wispy veils. "Comet" or "hairpin" inclusions formed from flux grains and droplets, rain"-like flux particles which can create a foggy appearance, straight and angular zoning. Kashan stones are noted for their high clarity.

Knischka

Ruby color	Tends to resemble that of natural Burmese rubies.
Inclusions	Two-phase negative crystals, large gas bubbles, small platinum platelets, net-like white flux fingerprints and wispy veils, ghost-like clouds, seed crystals, straight and angular color zoning, swirled color zones.

Ramaura

Fluorescence	May show small, chalky yellow areas under LW & SW. May show some slightly bluish white areas under SW. The overall fluorescent color is moderate to extremely strong, dull, chalky red to orangy red under LW and the same color but weak to strong intensity under SW.
Inclusions	Flux inclusions with a distinctive orange to yellow color and high relief, white or near colorless flux inclusions,planes, straight straie-like color zoning, V-shaped zoning planes, color swirls and streamers, fractures and healed fractures similar to those in natural rubies. Ramaura stones are noted for their high degree of transparency.

Flux overgrowth

Lechleitner

Description	A core (faceted or crystal "seed") of Verneuil or natural corundum coated with a thin layer of flux-growth corundum. Lechleitner stones come in several colors--red, blue, yellow, orange, pink, green, colorless, and color-change; and normally the core is Verneuil corundum.
Inclusions	White or near colorless flux fingerprints and wispy veils which crisscross and reduce transparency, gas bubbles and curved straie or color banding when core is of Verneuil corundum, straight and angular color zoning in the flux overgrowth, repeated twinning if natural cores are used.

Places of Origin and Some Corresponding Properties & Inclusions

Kashmir Sapphire

Fluorescence	Orangy fluorescence in certain areas, not uniform.

Inclusions	Very sharp, wide-spaced, straight zones, often with a chevron appearance (the closer the zones are the less likely it is to be Kashmir); powdery texture and glowy quality; feathery streamers like tiny pennants attached to strings; isolated short stubby needles; none of the standard rutile silk; fingerprints not common, included crystals are usually small clusters instead of large like in Ceylon material, devoid of a lot of inclusions compared to other localities.
Burma ruby	
Inclusions	Small nest-like concentrations of tiny rutile needles, color swirls, and streamers (but also present in Ramaura ruby), calcite and dolomite inclusions as well as spinel, corundum, and zircon; has a roiled (heavy graining) appearance, liquid-filled fingerprints and feathers tend to be absent.
Burmese sapphire	
Inclusions	Dense clouds of rutile silk similar to those in Burmese ruby, silk shorter and more densely packed though than in Sri Lankan stones; fingerprints common and typically look folded; exceptionally even color and no banding in most specimens; included crystals less common than in Burmese rubies.
Sri Lankan (Ceylon) ruby	
Inclusions	Very long, fine rutile needles that traverse the whole stone, numerous fingerprints, feathers, well-formed negative crystals, uneven coloring and large colorless areas, zircons surrounded by tension halos, other crystals such as calcite, garnet, pyrite, tourmaline, spinel and apatite.
Sri Lankan sapphire	
Fluorescence	Often a uniform orange or red fluorescence throughout the stone.
Inclusions	Generally the same as those in Ceylon ruby. Texture" clouds that are often brownish yellow may be present too.
Thai & Cambodian ruby	
Inclusions	"Saturn" inclusions which are negative or solid crystals surrounded by fingerprints, no rutile silk, wispy fingerprints resembling flux inclusions seen in synthetic corundum, crystals such as pyrrhotite or apatite or garnet, color zoning rare, repeated twinning common, often present are long boehmite needles which intersect in three directions almost at right angles to each other resembling a three-dimensional grid or jungle gym.

Kanchanaburi Thai sapphire

Fluorescence	Generally inert
Inclusions	Often slightly milky, no rutile silk, strong and uneven color zoning, long white boehmite needles, fingerprints, feathers, crystals such as feldspar and hornblende.

Chanthaburi & Trat Thai sapphire

Fluorescence	Generally inert
Inclusions	Small red and orange crystals which are often surrounded by small fingerprints with yellow stains, very sharp hexagonal growth zoning, slightly yellowish texture clouds, fingerprints, feathers, when present silk is usually found in narrow planes at the table and culet.

Cambodian sapphire

Fluorescence	Generally inert
Inclusions	White boehmite needles but no rutile silk, numerous fingerprints and feathers, atoll-like inclusions with crystal and halo, inclusions like those of Chanthaburi such as very sharp hexagonal color zoning and red uranium pyrochlore crystals (very characteristic of this sapphire).

Australian sapphire

Fluorescence	Generally inert
Dichroism	Very strong green to very dark violet-blue dichroism
Inclusions	Crystals (with or without halo) similar to those of Chanthaburi and Cambodia, fingerprints and feathers common, strong zoning and color banding, evenly colored stones may show sharp fine banding under magnification.

Tanzanian sapphire (Umba Valley)

Inclusions	Twinning planes and accompanying long boehmite needles, tiny thin plates or films perhaps of hematite, crystals of apatite.

Kenyan ruby

Fluorescence	(LW) strong to very strong red or red orange (SW) slightly weaker than long wave. It's strong fluorescence helps separate it from Thai ruby.

Inclusions	Numerous feathers and fingerprints (some resemble flux inclusions in synthetic corundum), white "texture" clouds, diffuse color zoning or sharp narrow color banding common, boehmite needles and twinning.

The information in this appendix is based mainly on the following sources.

Corundum by Richard Hughes

Gemmologists' Compendium by Robert Webster

GIA Gem Reference Guide

Handbook of Gem Identification by Richard Liddicoat

Photoatlas of Inclusions by Eduard Gubelin and John Koivula

Articles in *Gems & Gemology* listed in the bibliography.

Notes of the AGL Kashmir sapphire seminar at the Tucson 1990 gem show (C.R. Beesley, speaker).

Bibliography

Books

Ahrens, Joan & Malloy, Ruth. *Hong Kong Gems & Jewelry*. Hong Kong: Delta Dragon, 1986.

Anderson, B. W. *Gem Testing*. Verplanck, NY: Emerson Books, 1985.

Arem, Joel. *Gems & Jewelry*. New York: Bantam, 1986.

Australian Gem Industry Assn. *Australian Opals & Gemstones*. Sydney: Australian Gem Industry Assn, 1987.

Avery, James. *The Right Jewelry for You*. Austin, Texas: Eakin Press, 1988.

Babcock, Henry A. *Appraisal Principles and Procedures*. Washington DC: American Society of Appraisers, 1980.

Ball, Sydney H. *Roman Book on Precious Stones*. Los Angeles: G.I.A., 1950.

Bauer, Jaroslav & Bouska, Vladimir. *Pierres Precieuses et Pierres Fines*. Paris: Bordas, 1985.

Bauer, Dr. Max. *Precious Stones*. Rutland, Vermont & Tokyo: Charles E. Tuttle, 1969.

Bingham, Anne. *Buying Jewelry*. New York: McGraw Hill, 1989.

Bruton, Eric, *Legendary Gems or Gems that Made History*. Radnor, PA: Chilton 1986.

Ciprani, Curzio & Borelli, Alessandro. *Simon & Schuster's Guide to Gems and Precious Stones*. New York: Simon and Schuster, 1986.

Desautels, Paul E. *The Gem Kingdom*. New York: Random House.

Farrington, Oliver Cummings. *Gems and Gem Minerals*. Chicago: A. W. Mumford, 1903.

Federman, David & Hammid, Tino. *Consumer Guide to Colored Gemstones*. Shawnee Mission, KS: Modern Jeweler, 1989.

Fisher, P. J. *The Science of Gems*. New York: Charles Scribner's Sons, 1966.

Frank, Joan. *The Beauty of Jewelry*. Great Britain: Colour Library International, 1979.

Freeman, Michael. *Light*. New York: Amphoto, 1988.

Gemological Institute of America. *Gem Reference Guide*. Santa Monica, CA: GIA, 1988.

Goldemberg, Rose Leiman. *All About Jewelry*. New York: Arbor House, 1983.

Greenbaum, Walter W. *The Gemstone Identifier*. New York: Prentice Hall Press, 1988.

Grelick, Gary R. *Diamond, Ruby, Emerald, and Sapphire Facets*. Buffalo, NY: 1985.

Gubelin, Eduard J. *The Color Treasury of Gemstones*. New York: Thomas Y. Crowell, 1984.

Gubelin, Eduard J. & Koivula, John I. *Photoatlas of Inclusions in Gemstones*. Zurich: ABC Edition, 1986.

Hoskin, John & Lapin, Lindie. *The Siamese Ruby*. Bangkok: World Jewels Trade Centre, 1987.

Hughes, Richard W. *Corundum*. London: Butterworth-Heinemann, 1990

Jackson, Carole. *Color Me Beautiful*. New York: Ballantine, 1985.

Jewelers of America. *The Gemstone Enhancement Manual*. New York: Jewelers of America, 1990.

King, Dawn. *Did Your Jeweler Tell You?* Oasis, Nevada: King Enterprises, 1990.

Kraus, Edward H. & Slawson, Chester B. *Gems & Gem Minerals*. New York: McGraw-Hill, 1947.

Kunz, George Frederick. *The Curious Lore of Precious Stones*. New York: Bell, 1989.

Kunz, George Frederick. *Gems & Precious Stones of North America*. New York: Dover, 1968.

Kunz, George Frederick. *Rings for the Finger*. New York: Dover, 1917.

Liddicoat, Richard T. *Handbook of Gem Identification*. Santa Monica, CA: GIA, 1981.

Marcum, David. *Fine Gems and Jewelry.* Homewood, IL: Dow Jones-Irwin, 1986.

Matlins, Antoinette L. & Bonanno, A. *Gem Identification Made Easy.* South Woodstock, VT: Gemstone Press, 1989.

Matlins, Antoinette L. & Bonanno, A. *Jewelry & Gems the Buying Guide.* South Woodstock, VT: Gemstone Press, 1987.

Meen, V. B. & Tushingham, A. D. *Crown Jewels of Iran.* Toronto: University of Toronto Press, 1968.

Miguel, Jorge. *Jewelry, How to Create Your Image.* Dallas: Taylor Publishing, 1986.

Miller, Anna M. *Gems and Jewelry Appraising.* New York: Van Nostrand Reinhold Company, 1988.

Mumme, I. A. *The World of Sapphires.* Port Hacking, N.S.W.: Mumme Publications, 1988.

Nassau, Kurt. *Gemstone Enhancement*. London: Butterworths, 1984.

O'Donoghue, Michael. *Identifying Man-made Gems*. London: N.A.G. Press, 1983.

O'Neil, Paul. *Gemstones*. Alexandria, VA: Time-Life Books, 1983.

Parsons, Charles J. *Practical Gem Knowledge for the Amateur*. San Diego, CA: Lapidary Journal, 1969.

Pearl, Richard M. *American Gem Trails*. New York: McGraw-Hill, 1964.

Pough, Fredirick H. *The Story of Gems and Semiprecious Stones.* Irvington-on-Hudson, NY: Harvey House, 1967.

Preston, William S. *Guides for the Jewelry Industry.* New York: Jewelers Vigilance Committee, Inc., 1986.

Ramsey, John L. & Ramsey, Laura J. *The Collector/Investor Handbook of Gems.* San Diego, CA: Boa Vista Press, 1985.

Rubin, Howard. *Grading & Pricing with GemDialogue*. New York: GemDialogue Marketing Co., 1986.

Rutland, E. H. *An Introduction to the World's Gemstones.* Garden City, NY: Doubleday, 1974.

Schmetzer, Karl. *Naturliche und synthetische Rubine.* Stuttgart: E. Schweizerbart'sche Verlagsbuchhandlung, 1986.

Schumann, Walter. *Gemstones of the World.* New York: Sterling, 1977.

Sinkankas, John. *Gem Cutting: A Lapidary's Manual.* New York: Van Nostrand Reinhold, 1962.

Sinkankas, John. *Van Nostrand's Standard Catalogue of Gems.* New York: Van Nostrand Reinhold, 1968.

Webster, Robert. *Gemmologists' Compendium.* New York: Van Nostrand Reinhold, 1979.

Webster, Robert. *Practical Gemmology.* Ipswich, Suffolk: N. A. G. Press, 1976.

Weinstein, Michael. *Precious and Semi-Precious Stones.* London: Sir Isaac Pitman & Sons, 1946.

Wykoff, Gerald L. *Beyond the Glitter.* Washington DC: Adamas, 1982.

Zucker, Benjamin. *Gems & Jewels: A Connoisseur's Guide.* New York: Thames and Hudson, 1984.

Zucker, Benjamin. *How to Buy & Sell Gems: Everyone's Guide to Rubies, Sapphires, Emeralds & Diamonds.* New York: Times Books, 1979.

Articles

Allaman, John. "The new Kruss ultraviolet spectroscope UVS 2000 and its use in distinguishing between natural and synthetic rubies." *AGA Cornerstone*, July, pp. 25-27, 1990.

Altobelli, Cos. "Valuing diffusion-treated sapphires." *Jeweler's Circular Keystone*. October, pp. 46-50, 1990.

Arem, Joel E. "Ruby - The current market situation." *Precious Stones Newsletter.* Dec. pp. 6-9, 1980.

Atkinson, David & Kothavala, Rustamz. "Kashmir sapphire." *Gems & Gemology.* Vol. 19, No. 2, 1983.

Austin, Gordon T. "Montana gem production on the rise." *Colored Stone*. March/April, pp. 14 & 15, 1990.

Bancroft, Peter. "Rubies of Thailand." *Lapidary Journal.* October, pp. 45-55, 1988.

Beesley, C. R. "The alchemy of blue sapphire." *Jeweler's Circular Keystone*. August, pp. 102 & 103, 1982.

Beesley, C. R. "Detection of heated sapphire: Some tell-tale signs." *Jeweler's Circular Keystone*. August, pp. 106 & 107, 1982.

Beesley, C. R. "The Yehuda controversy, A laboratory perspective." *Modern Jeweler*. October, pp. 44-51, 1989.

Bender, Cynthia. "Truth or consequences, A legal guide for jewelers." *Modern Jeweler*. March, pp. 35-49, 1990.

Birmingham, Nan T. "Rare and royal rubies." *Town & Country*. June, pp. 117-119, 1980.

Carter, Virginia L. "The most common misconceptions about flux grown rubies." *Cornerstone*. July, pp. 39-41, 1990.

Chatham, Thomas. "Truth in appraising," *Cornerstone*. July, pp. 42 & 43, 1990.

Crowningshield, Robert. "Padparadscha: What's in a name?" *Gems & Gemology*. Vol. 19. No 1, pp. 30-35, 1983.

Crowningshield, Robert. "Gem Trade Lab Notes: Ruby simulants: Spinel and synthetic ruby doublet -- Sapphire: More colors of heat-treated stones." *Gems & Gemology*. Vol. 20, No. 4, pp. 231-232, 1984.

Crowningshield, Robert. "Gem Trade Lab Notes: Ruby, Natural ruby doublet -- Heated sapphire." *Gems & Gemology*. Vol. 23, No. 1, pp. 47-49, 1987.

Federman, David. "The treater's art: Triumph and trauma." *Modern Jeweler*. June, pp. 45-55, 1986.

Federman, David. "Look who's buying in Bangkok." *Modern Jeweler*. September, pp. 91-94, 1988.

Federman, David. "Last Word: Appraisal certification: The thunder grows nearer." *Modern Jeweler*. November, p. 99, 1988.

Federman, David. "Last Word: Suicidal silence." *Modern Jeweler*. July, p. 210, 1989.

Federman, David. "Last Word: Death by semantics." *Modern Jeweler*. November, p. 92, 1989.

Federman, David. "Lab-grown luxury." *Modern Jeweler*. November, pp. 46-52, 1989.

Federman, David. "Gem profile: Burma sapphire." *Modern Jeweler*. Dec., pp. 35-36, 1989.

Federman, David. "Last Word: Emerald oiling: Cause for celebration, not concern." *Modern Jeweler.* May, p. 94, 1990.

Frazier, Si & Ann. "Gemstones: South of the Equator." *Lapidary Journal.* August, pp. 36-39, 1990.

Fritsch, Emmanuel & Stockton, C. M. "Infrared spectroscopy in gem identification." *Gems & Gemology.* Vol. 23, No. 1, pp. 18-26, 1987.

Fritsch, Emmanuel & Rossman G. R. "An update on color in gems. Part 1: Introduction and colors caused by dispersed metal ions." *Gems & Gemology.* Vol. 23, No. 3, pp. 126-138, 1987.

Fritsch, Emmanuel & Rossman G. R. "An update on color in gems. Part 2: Colors involving multiple atoms and color centers." *Gems & Gemology.* Vol. 24, No. 1, pp. 3-14, 1988.

Fritsch, Emmanuel & Rossman G. R. "New technologies of the 1980's" *Gems & Gemology* Vol. 26, No. 1, pp. 64-75, 1990.

Furui, Wang. "The sapphires of Penglai, Hainan Island, China." *Gems & Gemology.* Vol. 24, No. 3, pp. 155-160, 1988.

Gubernick, Lisa. "You die for sure." *Forbes 400.* Oct. 26, 87, pp. 94-96.

Hanneman, W. W. "Gemological Instruments." *Lapidary Journal.* October pp. 74-82, 1989.

Hargett, David. "Gem Trade Lab Notes: Damaged "Burma" ruby." *Gems & Gemology.* Vol. 25, No. 4, pp. 240-241, 1989.

Hargett, David. "Gem Trade Lab Notes: Ruby, Verneuil synthetic with needle-like inclusions -- Sapphire, A large fine-color star." *Gems & Gemology.* Vol. 25, No. 1, pp. 38-39, 1989.

Hiss, Deborah A. "Fancy cuts, American style." *Jeweler's Circular Keystone.* May, pp. 166-170, 1988.

Hiss, Deborah A. "A jeweler's guide to gem instruments." *Jeweler's Circular Keystone.* August, pp. 480-488, 1988.

Hiss, Deborah A. "The jeweler's gemstone dilemma: It is natural, isn't it?" *Jeweler's Circular Keystone.* October, pp. 232-246, 1988.

Hiss, Deborah A. "Gemstones: What's going on in the gemstone market." *Jeweler's Circular Keystone.* August, p. 503 & 504, 1989.

Hiss, Deborah A. "Gem sources, prices, policies." *Jeweler's Circular Keystone.* August, pp. 510-518, 1989.

Hiss, Deborah A. "Synthetic colored gems: Breaking down barriers." *Jeweler's Circular Keystone*. September, Part II, pp. 65-72, 1989.

Hiss, Deborah A. & Holmes G. "Color enhancement: How to tell your customer." *Jeweler's Circular Keystone*. July, pp. 321-328, 1989.

Hiss, Deborah A. & Walowitz H. S. "Ruby: the lover's stone." *Jeweler's Circular Keystone*. June, pp. 181-188, 1988.

Hiss, Deborah A. & Walowitz H. S. "The many colors of sapphire." *Jeweler's Circular Keystone*. July, pp. 315-322, 1988.

Hofer, Stephen. "Colored diamonds." *New York Diamonds.* December, pp. 78-80, 1988.

Hofer, Stephen. "Colored diamonds." *New York Diamonds.* Autumn, pp. 70-72, 1989.

Hofer, Stephen. "Discussing color tone with confidence." *New York Diamonds*. Winter, pp. 68-72, 1990.

Hofer, Stephen. "Describing lightness and saturation." *New York Diamonds*. Spring, pp. 70-72. 1990.

Hofer, Stephen. "How angle of viewing affects perception." *New York Diamonds.* Summer, pp. 74-76, 1990.

Hofer, Stephen. "Interpreting face-up color." *New York Diamonds*. Autumn, pp. 70-72, 1990.

Hudson, Steve, "In the hands of the facetor it's the stuff of great art." *Rock & Gem*. May, pp. 9-12, 1990.

Kammerling, Robert C. "Red grow the rubies." *Jewelers Quarterly*. 3rd qtr, pp. 6-8 & 20, 1986.

Kammerling, Robert C., Koivula J. I., "GIA Focus: GIA spots diffusion-treated sapphires at Tucson Gem Show in February. *Jeweler's Circular Keystone*. May, pp. 46 & 48, 1990.

Kammerling, Robert C., Koivula J. I. "GIA Focus: Corundum doublets now prevalent." *Jeweler's Circular Keystone*. July, pp. 62 & 64, 1990.

Kammerling, Robert C., Koivula J. I. "Tips on identifying flame-fusion synthetics." *GIA Scope*. Spring/Summer, 1990.

Kammerling, Robert C., Koivula J. I. "Using magnification to detect assembled stones." *GIA Scope*. Fall, 1990.

Kammerling, Robert C., Koivula J. I. "Microscopic features of synthetic rubies Part 1: Melt Products" *GIA Scope*. Winter, 1990/1991.

Kammerling, Robert C., Koivula J. I., & Kane R. E. "Gemstone enhancement and its detection in the 1980's". *Gems & Gemology*. Vol. 26, No. 1, pp. 32-49, 1990.

Kane, Robert E. "The Ramaura synthetic ruby." *Gems & Gemology*. Vol. 19, No. 3, pp. 130-148, 1983.

Kane, Robert E. "Natural rubies with glass-filled cavities." *Gems & Gemology*. Vol. 20, No. 4, pp. 187-199, 1984.

Kane, Robert E. "A preliminary report on the new Lechleitner synthetic ruby and synthetic blue sapphire." *Gems & Gemology*. Vol. 21, No. 1, pp. 35-39, 1985.

Kane, Robert E. "Gem Trade Lab Notes: Synthetic brown star sapphire." *Gems & Gemology*, Vol. 24, No. 3, p. 173, 1988.

Kane, Robert E. "Gem Trade Lab Notes: Sapphire, pinkish orange ('Padparadscha')." *Gems & Gemology*. Vol. 22, No. 1, pp. 52-53, 1986.

Kane, Robert E. "Gem Trade Lab Notes: An unusual green star." *Gems & Gemology*. Vol. 25, No. 1, 1989.

Kane, Robert E., Kammerling, R. C., Koivula, J. I., Shigley, J. E., & Fritsch, E. "The identification of Blue Diffusion-Treated Sapphires." *Gems & Gemology*. Vol. 26, No. 2, pp. 115-133, 1990.

Keller, Alice S. & Keller P. "The Sapphires of Mingxi, Fujian Province, China." *Gems & Gemology*. Vol. 22, No. 1, pp. 41-45, 1986.

Keller, Peter C. "The rubies of Burma: A review of the Mogok stone tract." *Gems & Gemology*. Vol. 19, No. 4, pp. 209-219, 1983.

Keller, Peter C. & Fuquan W. "A survey of the gemstone resources of China." *Gems & Gemology*. Vol 22, No. 1, pp. 3-13, 1986.

Keller, Peter C., Koivula J. I., & Jara, G. "Sapphires from the Mercaderes--Rio Mayo area, Cauca, Colombia." *Gems & Gemology*. Vol. 21, No. 1, pp. 20-25, 1985.

Kiefert, Lore and Schmetzer, K. "Pink and violet sapphires from Nepal." *The Australian Gemmologist*, May pp. 225-230, 1987.

Koivula, John I. "Induced fingerprints." *Gems & Gemology*. Vol. 19, No. 4, pp. 220-227, 1983.

Koivula, John I. and Kammerling R. C. "A gemological look at Kyocera's new synthetic star ruby." *Gems & Gemology*. Vol. 24, No. 4, pp. 237-240, 1988.

Koivula, John I. and Kammerling R. C. "Gem News: New evidence of treatment in Umba sapphires." *Gems & Gemology*. Vol, 24, No. 4, p. 251, 1988.

Koivula, John I. and Kammerling R. C. "Gem News: Large bicolored sapphire, Kanchanaburi sapphires." *Gems & Gemology*. Vol. 25, No. 3, p. 181, 1989.

Koivula, John I. and Kammerling R. C. "Gem News: Diffusion treated corundum, New world sapphires, New synthetic ruby" *Gems & Gemology*. Vol. 26, No. 1, pp. 100-102 & 109, 1990.

Koivula, John I. and Kammerling R. C. "Production & identification of synthetic rubies." *Colored Stone*. May/June, pp. 9-12, 1990.

Koivula, John I. and Kammerling, R. C. "Gem News: Rubies from Vietnam." *Gems & Gemology*. Vol. 26, No. 2, pp. 163-164. 1990.

Lee, Tom L. "Gemstone enhancement...Then and Now." *Colored Stone. July/August, pp. 23-26, 1990.*

Liddicoat, Richard T. "What is a synthetic?" *Gems & Gemology*. Vol. 23, No. 3, p. 125, 1987.

Locke, Elizabeth. "The crown jewels of gemstones." *Town & Country*. June, pp. 154-159, 1990.

Martin, Deborah Dupont. "Gemstone durability: Design to display." *Gems & Gemology*. Vol. 23, No. 2, pp. 63-77, 1987.

Massimiano, Charise. "U.S. dealers mised over geuda 'grab-bag'" *Colored Stone*. Nov. & Dec., p. 12, 1990.

Nassau, Kurt. "Synthetic gem materials in the 1980's." *Gems & Gemology*. Vol. 26, No. 1, 1990.

Nassau, Kurt & Valente G.K. "The seven types of yellow sapphire and their stability to light." *Gems & Gemology*. Vol. 23, No. 4, pp. 222-231, 1987.

Pough, Frederick. "Corundum." *Lapidary Journal*. April, pp. 15-18, 1990.

Rubin, Howard. "The effects of lighting on gemstone colors." *Jewelers Quarterly*. 2nd quarter, p. 35, 1987.

Schmetzer, Karl & Medenbach O. "Examination of three-phase inclusions in colorless, yellow, and blue sapphires from Sri Lanka." *Gems & Gemology*. Vol. 24, No. 2, pp. 107-111, 1988.

Shigley, James E., Dirlam D. M., Schmetzer K., & Jobbins E. A. Gem localities of the 1980's" *Gems & Gemology*. Vol. 26, No. 1, pp. 4-29. 1990.

Shor, Russel. "Gemstones." *Jeweler's Circular Keystone*. December, pp. 134 & 135, 1988.

Shor, Russel & Reilley, B. "Gem Notes--Beesley: Treatment disclosure a must." *Jeweler's Circular Keystone*. September, pp. 48 & 49, 1990.

Smith, Jane V. "Ruby market going 'crazy.'" *Jeweler's Circular Keystone*. November, p. 168, 1987.

Themelis, Ted. "Inclusion of the month: Blue spot on ruby." *Lapidary Journal*. April, p. 19, 1988.

Themelis, Ted. "Inclusion of the month: Nepali corundum." *Lapidary Journal*. September, p. 19, 1989.

Themelis, Ted. "Clues to identity." *Lapidary Journal*. September, pp. 36-40, 1989.

Themelis, Ted. "Inclusion of the month: Crystals in ruby." *Lapidary Journal*. February, p. 19, 1990.

Themelis, Ted. "Inclusion of the month: Film in corundum." *Lapidary Journal*. October, p. 19, 1989.

Themelis, Ted. "Inclusion of the month: Some inclusions in Umba corundums." *Lapidary Journal*. November, p. 19, 1989.

Themelis, Ted. "Inclusion of the month: Oiling emerald." *Lapidary Journal*. March, p. 19, 1990.

Themelis, Ted. "Inclusion of the month: Yogo sapphire." *Lapidary Journal*. April, p. 19, 1990.

Themelis, Ted. "Oiling emeralds." *AGA Cornerstone*. July, pp. 21-24, 1990

Themelis, Ted & Federman, D. "A jeweler's guide to emerald oiling." *Modern Jeweler*. May, pp. 65-69, 1990.

Welch, Clayton. "Gem trade lab notes: Ruby, heat-treated natural and synthetic." *Gems & Gemology*. Vol. 23, No. 4, pp. 235-236, 1987.

Wickramanayake, D. "Gem Notes: Indian gem rush." *Jeweler's Circular Keystone*. November, p. 47, 1990.

Periodicals

Colored Stone. Devon, PA: *Lapidary Journal* Inc.

GAA Market Monitor Precious Gem Appraisal/Buying Guide. Pittsburgh, PA: GAA.

Gems and Gemology. Santa Monica, CA: Gemological Institute of America.

Gemstone Price Reports. Brussels: Ubige S.P.R.L.

The Guide. Chicago: Gemworld International, Inc.

Lapidary Journal. Devon, PA: Lapidary Journal Inc.

Jewelers Circular Keystone. Radnor, PA: Chilton Publishing Co.

Jewelers' Quarterly Magazine. Sonoma, CA.

Michelsen Gemstone Index. Pompano Beach, FL: Gem Spectrum.

Modern Jeweler. Lincolnshire, IL: Vance Publishing Inc.

National Jeweler. New York: Gralla Publications.

Palmieri's Auction/FMV Monitor. Pittsburgh, PA: GAA

Rock & Gem. Ventura, CA: Miller Magazines, Inc.

Miscellaneous: Courses, notes, and leaftlets

Gemological Institute of America Appraisal Seminar handbook.

Gemological Institute of America Gem Identification Course.

Gemological Institute of America Colored Stone Grading Course.

Gemological Institute of America Colored Stone Grading Course Charts, 1984 & 1989.

Gemological Institute of America Colored Stones Course. 1980 & 1989 editions.

Gemological Institute of America Jewelry Sales Course.

Notes of the AGL Kashmir sapphire seminar at the Tucson 1990 gem show (C.R. Beesley, speaker).

Tsavo Madini Inc. leaflet describing Tanzanian gems.

Index

Order Form

To: International Jewelry Publications
P.O. Box 13384
Los Angeles, CA 90013-0384 USA

Please send me:

___ copies of **THE RUBY & SAPPHIRE BUYING GUIDE.**

Within California $21.30 each (includes sales tax) ________

All other destinations $19.95 US each ________

___ copies of **THE DIAMOND RING BUYING GUIDE.**

Within California $13.82 each (includes sales tax) ________

All other destinations $12.95 US each ________

Postage & Handling for Books

USA: first book $1.50, each additional copy $.75 ________
Canada & foreign - surface mail: first book $2.50, ea. addl. $1.50 ________
Canada & Mexico - airmail: first book $3.75, ea. addl. $2.50 ________
All other foreign destinations - airmail: first book $9.00, ea. addl. $5.00 ________

___ copies of **DIAMONDS: FASCINATING FACTS.**

Within California $4.22 each (includes sales tax) ________

All other destinations $3.95 US each ________

Postage for Diamonds: Fascinating facts
USA: $0.55 per booklet ________
Canada & Mexico - airmail: $0.80 per booklet ________
All other foreign destinations - airmail: $1.25 per booklet ________

Total Amount Enclosed ________
(USA funds drawn on a USA bank)

Ship to:

Name________________________________

Address________________________________

City________________ State or Province________________

Postal or Zip Code________________ Country________________

OTHER PUBLICATIONS BY RENEE NEWMAN

The Diamond Ring Buying Guide: How to Spot Value & Avoid Ripoffs

A comprehensive guide to evaluating, selecting, pricing, and caring for diamond jewelry.

Discover:

- How to judge diamond quality and jewelry craftsmanship
- How to detect diamond and gold imitations
- How to choose between platinum, white gold, and yellow gold
- How to select a ring style that's both practical and flattering
- How to compare the prices of diamonds and jewelry mountings

"**Will definitely help consumers** . . . Written in a popular style with lots of personalized examples, the book should be easy reading for the young people who are thinking about their first diamond purchases."
Lapidary Journal

"**A wealth of information** . . . delves into the intricacies of shape, carat weight, color, clarity, setting style, and cut--happily avoiding all industry jargon and keeping explanations streamlined enough so even the first-time diamond buyer can confidently choose a gem."
American Library Association's *Booklist*

"There's no other book that gives as much detailed information on diamond jewelry pricing or as much **practical advice on selecting diamonds and ring mountings."**
Antique Showcase

"**A fact-filled text devoid of a lot of technical mumbo-jumbo.** Sellers . . . might do themselves a favor by making this book available to those clients who are eager to find that optimum blend of value, quality and service. This is a definite thumbs up!"
C. R. Beesley, President, American Gemological Laboratories, New York.
Jewelers' Circular Keystone

AVAILABLE AT bookstores, jewelry supply stores, rock shops, the GIA, and through the *Lapidary Journal* & Jeweler's Book Clubs, or by mail: See reverse side for order form.

151 pages, 85 black and white photos, 7" by 9", $12.95 US.

Diamonds: Fascinating Facts

An informative booklet with entertaining facts, poems, and statistics about diamonds.

A novel and appropriate greeting card to include with a diamond gift. It comes with a 6" x 9" white envelope. The inside front cover is designed to allow for a personal message.

Full-color, 16-page, self-cover booklet with six 5" x 7 1/2" photos; $3.95 US.

Order Form

To: International Jewelry Publications
P.O. Box 13384
Los Angeles, CA 90013-0384 USA

Please send me:

___ copies of **THE RUBY & SAPPHIRE BUYING GUIDE.**

Within California $21.30 each (includes sales tax) ________

All other destinations $19.95 US each ________

___ copies of **THE DIAMOND RING BUYING GUIDE.**

Within California $13.82 each (includes sales tax) ________

All other destinations $12.95 US each ________

Postage & Handling for Books

USA: first book $1.50, each additional copy $.75 ________
Canada & foreign - surface mail: first book $2.50, ea. addl. $1.50 ________
Canada & Mexico - airmail: first book $3.75, ea. addl. $2.50 ________
All other forcign destinations - airmail: first book $9.00, ea. addl. $5.00 ________

___ copies of **DIAMONDS: FASCINATING FACTS.**

Within California $4.22 each (includes sales tax) ________

All other destinations $3.95 US each ________

Postage for Diamonds: Fascinating facts
USA: $0.55 per booklet ________
Canada & Mexico - airmail: $0.80 per booklet ________
All other foreign destinations - airmail: $1.25 per booklet ________

Total Amount Enclosed ________
(USA funds drawn on a USA bank)

Ship to:

Name________________________________

Address________________________________

City________________ State or Province________________

Postal or Zip Code____________ Country________________

OTHER PUBLICATIONS BY RENEE NEWMAN

The Diamond Ring Buying Guide: How to Spot Value & Avoid Ripoffs

A comprehensive guide to evaluating, selecting, pricing, and caring for diamond jewelry.

Discover:

- How to judge diamond quality and jewelry craftsmanship
- How to detect diamond and gold imitations
- How to choose between platinum, white gold, and yellow gold
- How to select a ring style that's both practical and flattering
- How to compare the prices of diamonds and jewelry mountings

"**Will definitely help consumers** . . . Written in a popular style with lots of personalized examples, the book should be easy reading for the young people who are thinking about their first diamond purchases."
Lapidary Journal

"**A wealth of information** . . . delves into the intricacies of shape, carat weight, color, clarity, setting style, and cut--happily avoiding all industry jargon and keeping explanations streamlined enough so even the first-time diamond buyer can confidently choose a gem."
American Library Association's *Booklist*

"There's no other book that gives as much detailed information on diamond jewelry pricing or as much **practical advice on selecting diamonds and ring mountings."**
Antique Showcase

"**A fact-filled text devoid of a lot of technical mumbo-jumbo.** Sellers . . . might do themselves a favor by making this book available to those clients who are eager to find that optimum blend of value, quality and service. This is a definite thumbs up!"
C. R. Beesley, President, American Gemological Laboratories, New York.
Jewelers' Circular Keystone

AVAILABLE AT bookstores, jewelry supply stores, rock shops, the GIA, and through the *Lapidary Journal* & Jeweler's Book Clubs, or by mail: See reverse side for order form.

151 pages, 85 black and white photos, 7" by 9", $12.95 US.

Diamonds: Fascinating Facts

An informative booklet with entertaining facts, poems, and statistics about diamonds.

A novel and appropriate greeting card to include with a diamond gift. It comes with a 6" x 9" white envelope. The inside front cover is designed to allow for a personal message.

Full-color, 16-page, self-cover booklet with six 5" x 7 1/2" photos; $3.95 US.